Pandemic Detection and Analysis Through Smart Computing Technologies

Pandemic Detection and Analysis Through Smart Computing Technologies

Edited by
Ram Shringar Raw, PhD
Vishal Jain, PhD
Sanjoy Das, PhD
Meenakshi Sharma, PhD

First edition published 2022

Apple Academic Press Inc.
1265 Goldenrod Circle, NE,
Palm Bay, FL 32905 USA
4164 Lakeshore Road, Burlington,
ON, L7L 1A4 Canada

CRC Press
6000 Broken Sound Parkway NW,
Suite 300, Boca Raton, FL 33487-2742 USA
2 Park Square, Milton Park,
Abingdon, Oxon, OX14 4RN UK

© 2022 Apple Academic Press, Inc.

Apple Academic Press exclusively co-publishes with CRC Press, an imprint of Taylor & Francis Group, LLC

Library and Archives Canada Cataloguing in Publication

Title: Pandemic detection and analysis through smart computing technologies / edited by Ram Shringar Raw, PhD, Vishal Jain, PhD, Sanjoy Das, PhD, Meenakshi Sharma, PhD.

Names: Raw, Ram Shringar (Lecturer in computer science), editor. | Jain, Vishal, 1983- editor. | Das, Sanjoy, 1979- editor. | Sharma, Meenakshi (Lecturer in computer science), editor.

Description: First edition. | Includes bibliographical references and index.

Identifiers: Canadiana (print) 20210393513 | Canadiana (ebook) 20210393653 | ISBN 9781774910320 (hardcover) | ISBN 9781774910337 (softcover) | ISBN 9781003281610 (ebook)

Subjects: LCSH: COVID-19 (Disease)—Epidemiology—Data processing. | LCSH: COVID-19 (Disease)—Early detection—Data processing. | LCSH: Artificial intelligence—Medical applications. | LCSH: Internet of things.

Classification: LCC RA644.C67 P36 2022 | DDC 614.5/924140285—dc23

Library of Congress Cataloging-in-Publication Data

....

CIP data on file with US Library of Congress

....

ISBN: 978-1-77491-032-0 (hbk)
ISBN: 978-1-77491-033-7 (pbk)
ISBN: 978-1-00328-161-0 (ebk)

About the Editors

Ram Shringar Raw, PhD
*Associate Professor, Department of Computer Science and Engineering,
Netaji Subhas University of Technology, Delhi, India*

Ram Shringar Raw, PhD, is working in the Department of Computer Science and Engineering of Netaji Subhas University of Technology, East Campus, Delhi, India. He formerly worked as an Associate Professor in the Department of Computer Science, Indira Gandhi National Tribal University (A Central University, M.P.). He has more than 18 years of teaching, administrative, and research experience. Currently, he is associated with a wide range of journals and conferences as chief editor, editor, chair, and member. He has guided many MTech and PhD students for their dissertation and thesis. Currently, he is guiding two MTech and four PhD students. His current research interest includes mobile ad hoc networks, vehicular ad hoc networks, and IoT cloud-based networks. Dr. Raw has published two books and more than 100 research papers in international journals and conferences, including those from IEEE, Elsevier. Springer, Wiley & Sons, Taylor & Francis, Inderscience, Hindawi, IERI Letters, American Institute of Physics, etc.

Dr. Raw has received his BE (Computer Science and Engineering) and MTech (Information Technology) degrees in 2000 and 2005, respectively. He has obtained his PhD (Computer Science and Technology) from the School of Computer and Systems Sciences, Jawaharlal Nehru University, New Delhi, India, in 2011.

Vishal Jain, PhD
*Associate Professor, Department of Computer Science and Engineering,
School of Engineering and Technology, Sharda University,
Greater Noida, UP, India*

Vishal Jain, PhD, is an Associate Professor in the Department of Computer Science and Engineering, School of Engineering and Technology, Sharda University, Greater Noida, U.P., India. Before that, he has worked for several years as an Associate Professor at Bharati Vidyapeeth's Institute of Computer Applications and Management (BVICAM), New Delhi. He has more than 14 years of experience in the academics. He has more than 400 research citation indices with Google Scholar (h-index score 10 and i-10 index 11). He has authored more than 85 research papers in conferences and journals, including the Web of Science and Scopus. He has authored and edited more than 10 books with various reputed publishers, including Springer, Apple Academic Press, CRC, Taylor and Francis Group, Scrivener, Wiley, Emerald, and IGI-Global. His research areas include information retrieval, semantic web, ontology engineering, data mining, ad hoc networks, and sensor networks. He received a Young Active Member Award for the year 2012–13 from the Computer Society of India and a Best Faculty Award for the year 2017 and Best Researcher Award for the year 2019 from BVICAM, New Delhi. He holds PhD (CSE), MTech (CSE), MBA (HR), MCA, MCP, and CCNA.

Sanjoy Das, PhD
Associate Professor and Head, Department of Computer Science, Indira Gandhi National Tribal University, Amarkantak, M.P., India

Sanjoy Das, PhD, is an Associate Professor and Head of the Department of Computer Science, Indira Gandhi National Tribal University (a Central Government University), Amarkantak, M.P. (Manipur Campus), India. Before joining IGNTU he worked as an Associate Professor at the School of Computing Science and Engineering, Galgotias University, India, as well as Assistant Professor. He was also an Assistant Professor at G. B. Pant Engineering College, Uttarakhand, and Assam University, Silchar, India. His current research interests include mobile ad hoc networks and vehicular ad hoc networks, distributed systems, and data mining. He has published numerous papers in international journals and conferences, including those associated with IEEE and Springer. He did his BE, MTech, and PhD degrees in Computer Science.

Meenakshi Sharma, PhD
Galgotia University, Greater Noida, India

Meenakshi Sharma, PhD, is a Dean and Professor in the School of Computer Science & Engineering of the University Center of Research & Development (UCRD) at Galgotias University, Greater Noida, India. She has over 16 years of experience in teaching and research. She is a Senior Member of IEEE and is a highly qualified professional with an MTech in Computer Science & Engineering and a PhD in Computer Science (both from Kurukshetra University). Her research interests are machine learning, image processing, big data analytics, data compression, and digital and data warehousing. She has published over 60 research papers in IEEE Transaction, SCIE, SCI and Scopus in machine learning, deep learning, and AI, in collaboration with international authors. She has published four international grant patents and seven Indian patents. She was awarded a Best Research and Teacher Award in 2017 and 2018. Dr. Sharma has capably guided three PhD candidates and 40+ students in undergraduate/postgraduate programs, with five currently under guidance. She is valuable member of various engineering societies, including ISTE, ACM, InSc, ISDS Society, Japan, IEAE, and many others.

Contents

Contributors

Khalid Alfatmi
Assistant Professor, Department of Computer Engineering, SVKM's Institute of Technology, Dhule, Maharashtra, India

Basudeba Behera
Department of Electronics and Communication Engineering, National Institute of Technology Jamshedpur, Jharkhand, India, Tel.: +91-8812016250, E-mail: basudeb.ece@nitjsr.ac.in

Anjali Chaudhary
Assistant Professor-CSE, Greater Noida Institute of Technology (GNIOT), Plot-7, Knowledge Park II, Greater Noida, Uttar Pradesh – 201310, India, E-mail: anjali.chaudhary22@gmail.com

Sandip T. Chavan
School of Mechanical Engineering, Dr. Vishwanath Karad MIT World Peace University, Pune, Maharashtra, India, E-mail: sandip.chavan@mitwpu.edu.in

Colin E. Evans
Program for Lung and Vascular Biology, Stanley Manne Children's Research Institute, Ann and Robert H. Lurie Children's Hospital of Chicago, Chicago, IL, USA; Department of Pediatrics, Division of Critical Care, Northwestern University Feinberg School of Medicine, Chicago, IL, USA

Neha Gupta
Department of Biosciences, Faculty of Natural Sciences, Jamia Millia Islamia, New Delhi – 110025, India, E-mail: nehaguptalifesciences@gmail.com

Ujjwal Gupta
Department of Electronics and Communication Engineering, National Institute of Technology Jamshedpur, Jharkhand, India

Prachi Joshi
Assistant Professor, Department of Information Technology, Almora Campus, Soban Singh Jeena University, Almora, Uttarakhand, India, E-mail: joshiprachi068@gmail.com

K. Kalaiselvi
Professor and Head, Department of Computer Science, School of Computing, Vels Institute of Science, Technology, and Advanced Studies (Formerly Vels University), Chennai, Tamil Nadu, India, E-mail: kalairaghu.scs@velsuniv.ac.in

D. Karthika
Assistant Professor, Department of Computer Science, Chrompet, Chennai. Email: Karthika.d@sdnbvc.edu.in

Sapna Kataria
Assistant Professor-CSE, Noida International University, Plot-1, Sector-17A, Yamuna Expressway, Uttar Pradesh – 203201, India, E-mail: sapna.kataria@niu.edu.in

Mayuri Diwakar Kulkarni
Assistant Professor, Department of Computer Engineering, SVKM's Institute of Technology, Dhule, Maharashtra, India, E-mail: mayuridkulkarni@gmail.com

Bhagwati Prasad Pande
Assistant Professor, Department of Computer Applications, LSM Government PG College,
Pithoragarh, Uttarakhand, India, E-mail: bp.pande21@gmail.com

Rajesh V. Patil
Assistant Professor, School of Mechanical Engineering, Dr. Vishwanath Karad MIT World Peace
University, Pune – 411038, Maharashtra, India**, E-mail: patilraje@gmail.com**

Sagar Rai
Department of Electronics and Communication Engineering, National Institute of Technology
Jamshedpur, Jharkhand, India

Jayachandran Kizhakoot Ramachandran
HCL Technologies Ltd. Noida, Uttar Pradesh, India, E-mail: jayachandran.ki@hcl.com

Puneet Sachdeva
HCL Technologies Ltd. Noida, Uttar Pradesh, India, E-mail: sachdeva-p@hcl.com

Mohd. Faiz Saifi
Department of Biosciences, Faculty of Natural Sciences, Jamia Millia Islamia, New Delhi – 110025,
India

Neeta Sharma
Head of the Department-CSE, Noida International University, Plot-1, Sector-17A,
Yamuna Expressway, Uttar Pradesh – 203201, India, E-mail: neeta.sharma@niu.edu.in

Varsha Singh
Centre for Life Sciences, Chitkara School of Health Sciences, Chitkara University, Punjab, India,
E-mail: varsha.singh@chitkara.edu.in

Vinamrita Singh
Netaji Subhas University of Technology (Formerly Ambedkar Institute of Advanced Communication
Technologies and Research, Government of NCT of Delhi), Delhi, India,
E-mails: drvinamrita@aiactr.ac.in; vinamritasingh.phy@gmail.com

Abhishek M. Thote
Assistant Professor, School of Mechanical Engineering, Dr. Vishwanath Karad MIT World Peace
University, Pune – 411038, Maharashtra, India, Phone: +91-8446640525,
E-mail: abhi.thote8@gmail.com

Satishkumar Varma
Professor, HOD of Information Technology Department, Pillai College of Engineering (Affiliated to
the University of Mumbai), New Panvel, Navi Mumbai, Maharashtra – 410206, India,
E-mail: vsat2k@mes.ac.in

Abhishek A. Vichare
PhD Scholar, Department of Computer Engineering, Pillai College of Engineering (Affiliated to the
University of Mumbai), New Panvel, Navi Mumbai, Maharashtra – 410206, India,
E-mail: vichare1@gmail.com

Abbreviations

ABS	acrylonitrile butadiene styrene
AGMPs	aerosols generating medical procedures
AI	artificial intelligence
AlGaN	aluminum gallium nitride
ANN	artificial neural network
APH	augmented personalized healthcare
API	application programming interfaces
ARDS	advanced respiratory distress syndrome
ARIMA	autoregressive integrated moving average
ATR-FTIR	attenuated total reflection Fourier-transform infrared
BIS	behavioral immune system
BLE	Bluetooth low energy
BPL	below poverty level
CAT	computerized axial tomography
CC	cognitive computing
CCPA	California Consumer Privacy Act
CDC	center for disease control
CI	calculated imaging
CMV	cytomegalovirus
CNNs	convolution neural network
COAI	cellular operators association of India
COVID-19	coronavirus disease 2019
CoVs	coronavirus
CSP	cloud service providers
CT	computed tomography
CVD	cardiovascular diseases
CW	continuous wave
DBA	database administrator
DBMS	database management system
DLP	digital light processing
DoT	Department of Telecommunication
DRISHTI	digital real-time artificial intelligence system for social distancing with timely intervention

DST	Department of Science and Technology
EA	evolutionary algorithms
ECDPC	European Center for Disease Prevention and Control
EMRs	e-medical registers
EPA	Environmental Protection Agency
FDA	Food and Drug Administration
FDM	fused deposition modeling
FET	field-effect transistor
FTIR	Fourier transformed infrared spectroscopy
GA	genetic algorithm
GAD-7	generalized anxiety disorder
GASMAS	gas in scattering media absorption spectroscopy
GDP	gross domestic product
GDPR	general data protection regulation
GIS	geographical information system
GPMB	global preparedness monitoring board
GPS	global positioning system
HD	high definition
HPC	high-performance computing
HR	human resources
IATA	International Air Transport Association
ICTV	International Committee on Taxonomy of Viruses
IES-R	effect of event scale-revised
IMF	International Monetary Fund
IoHT	internet of healthcare things
IoMT	internet of medical things
IoT	internet of things
IR	infrared
ISP	Internet service provider
IT	information technology
JHU CSSE	Johns Hopkins University Center for Systems Science and Engineering
LAN	local area network
LDA	linear discriminant analysis
LEDs	light-emitting diodes
LSPR	localized surface plasmon resonance
MDD	major distressing disorder
MeitY	Ministry of Electronics and Information Technology

MERS	Middle East respiratory syndrome
MHA	Ministry of Home Affairs
MHFW	Ministry of Health and Family Welfare
MHRD	Ministry of Human Resource Development
ML	machine learning
MSI	medical services internet
NCDs	non-communicable diseases
nCoV	novel coronavirus
NIR	near-infrared
NLP	natural language processing
NLP	neuro-linguistic programming
NRAI	National Restaurant Association of India
OTP	one time password
PC	perceptual computing
PCA	principal component analysis
PCR	polymerase chain reaction
PDMS	polydimethylsiloxane
PGHD	patient health data
PHQ	patient health questionnaire
PII	personally identifiable information
PMMA	poly(methyl methacrylate)
PPE	personal protective equipment
PPT	plasmonic photothermal
PVC	polyvinyl chloride
QDA	quadratic discriminant analysis
RDBMS	relational DBMS
RDS	relational database service
RNA	ribonucleic acid
RSSI	received signal strength indication
RT-PCR	reverse transcription-polymerase chain reaction
SARS	severe acute respiratory syndrome
SARS-CoV	severe acute respiratory syndrome-corona virus
SARS-CoV-2	severe acute respiratory syndrome coronavirus 2
SAS	self-rating anxiety scale
SD	standard definition
SDS	self-rating depression scale
SFV	spumavirus
SLA	stereolithography

SLS/SLM	selective laser sintering/melting
SMS	short-messaging services
SPA	successive projections algorithm
SQL	structured query language
SSN	syndromic surveillance networks
STDs	sexually transmitted diseases
TIR	total internal reflection
UAV	unmanned aerial vehicles
UNCTAD	United Nation Conference on Trade and Development
UNESCO	United Nations Educational, Scientific, and Cultural Organization
US	United States
UTM	universal transport medium
UV	ultraviolet
UVGI	UV germicidal irradiation
WBE	wastewater-based epidemiology
WHO	World Health Organization

Preface

The world is now facing the most lethal pandemic, COVID-19. Since the pandemic outbreak, it collapsed human lives, economy, etc., and paralyzed medical science. Day by day, the number of infections among people is rapidly increasing as well as death. The world comes to a complete halt due to virtual and physical lockdown. This pandemic outbreak has created a global health and economic crisis after a few months. COVID-19 has influenced the way that we perceive our world and our everyday lives. This pandemic highly spreads with the possibility of causing severe respiratory disease. It has rapidly failed various government health policies, public health organizations, industries, and declared a global health emergency. Due to the COVID-19, millions of lives have been significantly altered, and demanding stress-free adjustment process is ongoing. Preventing and optimizing the rapid spreading of a pandemic is a big challenge due to the large population in the world and massive crowded cities. On COVID-19, how can different fields of science and technology, research, and laboratory give novel concept, investigations, theoretical, and practical views to foster positive health attitudes and adherence to preventive measures?

Many organizations in the world have invited and opened online portals to provide preventive measures and psychological counseling services on the COVID-19 pandemic for globally affected people. In addition, as part of interdisciplinary research topics, there may be a lot of theoretical reviews, research papers, and practical studies that have been published on different pandemics worldwide. The computing and communication technologies have been advanced day by day and very helpful in managing this pandemic in different means. The advancements in computational and computer science include various domains like machine learning, big data, internet of things (IoT), cloud computing, data science, artificial intelligence (AI), etc. Our target is to analyze various issues related to computing, simulate novel investigations and how they will be beneficial in an insight understanding of the pandemic. How technological advancement be helpful in curving the pandemic so that society is affected less. During this COVID-19 pandemic which is a very interesting and exciting situation when we can expect the possible outcome as a chapter for a book from different fields of computer science and other domains. This book is ideally designed for data scientists, doctors, engineers, economists, and many other professionals. This book will be very much helpful to the UG/PG and research scholar to understand the pre and post-effect of this pandemic.

CHAPTER 1

Modes of Transmission of Coronavirus

MOHD. FAIZ SAIFI,[1] COLIN E. EVANS,[2,3] and NEHA GUPTA[1]

[1]*Department of Biosciences, Faculty of Natural Sciences, Jamia Millia Islamia, New Delhi – 110025, India, E-mail: nehaguptalifesciences@gmail.com (N. Gupta)*

[2]*Program for Lung and Vascular Biology, Stanley Manne Children's Research Institute, Ann and Robert H. Lurie Children's Hospital of Chicago, Chicago, IL, USA*

[3]*Department of Pediatrics, Division of Critical Care, Northwestern University Feinberg School of Medicine, Chicago, IL, USA*

ABSTRACT

Transmission of the pathogen plays a central role in disease biology and epidemiology. For viral infections, transmission is known to occur when the pathogen (virion) leaves its reservoir or host through a portal of exit and is then carried further by some mode of transmission and then via an appropriate entry portal infects the susceptible host. The newly emerging viral diseases are major threats to public health. Contemporary pandemics such as SARS and COVID-19 occur by the emergence of highly transmissible and pathogenic coronaviruses viz. SARS-CoV and SARS-CoV-2, respectively. The World Health Organization has declared a worldwide emergency over the recent widespread of SARS-CoV-2, mediated infection among humans, widely known as COVID-19, a pandemic that might result in human death. The exact source of SARS-CoV-2 is still unknown; however, the disease might spread through COVID-19-positive individuals or a contaminated environment. Currently available information indicates that both SARS-CoV and SARS-CoV-2 might have originated from their common natural host bat and transmitted the infection to humans via some intermediate animal host possibly through the mechanism of cross-species transmission. Hence, understanding various modes of transmission may prove beneficial in designing the strategies to break the chain of infection. The current chapter, thus describes the various possible modes of transmission for the emergent SARS-CoV-2 with a discussion on the mechanism of cross-species transmission and factors associated with the spread of SARS-CoV and SARS-CoV-2.

1.1   INTRODUCTION

The pathogen's transmission is crucial to disease biology and epidemiology. Transmission of viral infections is believed to occur when the pathogen (virion) exits its reservoir or host through an exit portal [1] and is then carried further by some mode of transmission and then via an appropriate entry portal infects the susceptible host. The majority of newly emerging viral diseases are becoming a public health threat [1]. Particularly, viruses emerging from wildlife hosts become a cause of high-impact disease as found in SARS, Ebola fever, influenza, and lately emerging COVID-19 in humans. The transfer of infectious agents from animals to humans via the mechanism of cross-species transmission is primarily responsible for the development of these diseases [2]. Furthermore, the transmission of the infectious agent can occur either as a vertical disease agent or a horizontal disease agent [3]. Vertical disease transmission involves the transmission of pathogens from parent to offspring such as in prenatal transmission while Horizontal transmission involves the transmission of pathogens from one individual to another in the same generation through various ways including physical contact, contaminated food, body fluids, airborne inhalation or via vectors [3]. Pandemics such as SARS and COVID-19 are considered to occur as a result of the advent of highly transmissible pathogens and pathogenic coronaviruses (CoVs) viz. SARS-CoV and SARS-CoV-2, respectively [3]. These coronaviruses cause infections of the lower respiratory tract, and SARS-CoV was first identified in Guangdong province, China, in 2002, and later in Wuhan, China, in December 2019 as SARS-CoV-2 [1]. Both SARS-CoV and SARS-CoV-2 are known to be originated from their common natural host bat and then transmitted the infection to humans via some intermediate animal host possibly through the mechanism of cross-species transmission. The current chapter describes the various possible modes of transmission for the emergent SARS-CoV-2 with a discussion on the mechanism of cross-species transmission and factors associated with the spread of SARS-CoV and SARS-CoV-2 [3]. Information provided here shall provide an in-depth understanding of various factors that are responsible for different modes of transmission for CoVs among humans and animals and may prove beneficial towards designing the strategies to break the chains of infection [4].

1.2   GENERAL ASPECTS RELATED TO TRANSMISSION

A global pandemic is characterized by the widespread emergence of chains of Infection. Modes of Transmission form the indispensable component of these chains of infection. To understand the complete picture of the emergence of infectious disease, we must know about the interaction of the infectious agent with the host and the environment. Infectious disease is caused by the interaction of pathogenic agent, host, and environment, according to the traditional epidemiologic triad model. These interactions are composed of several components and processes, which are often described using key terminologies. Figure 1.1 gives a diagrammatic representation of the traditional epidemiologic triad model. The current section defines the commonly used terminologies used to describe 'Chains of infections' as follows:

1. **Transmission:** The pathogen moves from the source of infection (individual or group) to a new host through this process [1].
2. **Host:** A host is an organism that bears mutualistic, parasitic, and commensal (Symbiont) guest (pathogen or causative agent) [2]. The host usually provides the guest with food and shelter. The simplest examples are animals serve as hosts to parasitic worms; a virus replicates itself in an organism or cell culture. Further, the host may also be classified as a Definitive (Primary) Host, which is an organism in which the parasite matures and reproduces sexually or secondary (intermediate host)—an organism that carries a sexually immature parasite and is needed by the parasite to complete its lifecycle.

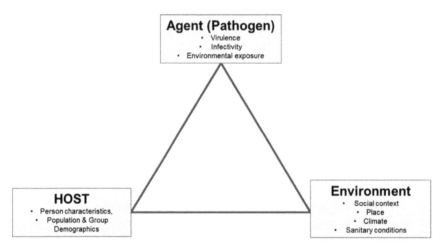

FIGURE 1.1 Traditional 'epidemiologic triad' model.

To be able to survive for an extended period of time, pathogens (microorganisms) require reservoirs. Thus, the reservoir is the habitat where a pathogen obtains nutrition, grows, multiplies, and resides alive. Reservoirs can be either be living or non-living sites [3]. A reservoir, it should be noted, may or may not be the source from which a pathogen is transmitted to the host. For example, soil serves as a reservoir for *Clostridium botulinum*, but inappropriately canned food is the source of most infections in this case. Humans, too, act as reservoirs for a number of infectious diseases [4]. In sexually transmitted diseases (STDs), mumps, measles pathogens are transmitted from one person to another without the use of an intermediate host [5]. The smallpox virus, for which humans were the only reservoir, was declared eradicated after the last human case was detected and isolated [6]. When a pathogen leaves its reservoir then after the completion of the life cycle into the host, it causes a communicable disease that point is known as the portal of exit [7]. Thus, in general terms, the route through which the pathogen exits the host, which typically corresponds to the pathogen's location, is referred to as the portal of exit. As an example, influenza viruses exit the respiratory tract, *cholera vibrios* in feces, blood-borne agents exit either by crossing the placenta from mother to the fetus or through cuts, needles in the skin, or blood-sucking arthropods [8]. The term "portal of entry" refers to the "means" by which a pathogen enters a susceptible host. The portal of entry provides access to the tissues/sites where the pathogen can multiply. Infectious agents also use the same gateway to reach a new host as they did to leave the previous host [9]. For example, the influenza virus exits the respiratory tract of the source host and enters the new host through the respiratory tract only. Many pathogens that cause gastroenteritis, on the other hand, take a fecal-oral route, in which they leave the source host in feces and are carried away by unwashed hands or a vehicle such as water or food, and then enter the host via ingestion through the mouth [10]. Thus, a pathogen exits the reservoir via a portal of exit, is transmitted via an appropriate mode of transmission, and reaches the reservoir via an appropriate portal of entry., the pathogen gets entry into the susceptible host or cause infection and this whole series of infection or a pathogen's transmission is referred to as a "chain of infection" [11]. Further, Figure 1.2 represents the sequence involved in the transmission of a pathogen to produce a chain of infection.

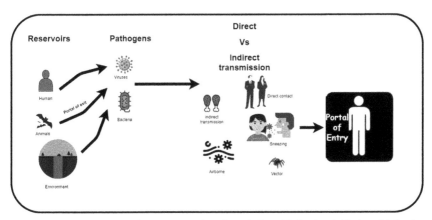

FIGURE 1.2   Diagrammatic representation for 'chain of infection.' Chain of infection comprises of three main parts-reservoir, modes of transmission and susceptible host.

Pathogens can also use a dormancy mechanism, in which they live (but do not reproduce) in non-living environments for varying periods of time; for example, *Clostridium tetani* may survive in the soil and as a resistant endospore [12]. Viruses such as the influenza virus, which, after varying amounts of time on a banknote, can infect a cell culture and can live for 48 hours to 17 days. After varying amounts of time on a banknote, it can infect a cell culture, depending on the manner in which they are deposited on the banknote. Thus, an individual without showing any symptoms can transmit the pathogen to another and is known as 'Carrier'. There are different kinds of carriers present in nature; a passive carrier is an organism already contaminated with the pathogen and has a capacity to transfer the pathogen mechanically, however, it is not infected [13]. A doctor or other health care worker who did not wash their hands after seeing a patient with infectious agents, for example, might become a passive carrier of the pathogen and infect another patient. In contrast, an active carrier is a person (host) who is already infected and can transmit the pathogen to others. It may or may not display the symptoms of infection [14]. Even, an active carrier can transfer the pathogen or infection during the incubation period (incubation period is the period before the infected person shows any symptoms or signs of infection). Besides, a person (host) who has clinically recovered from disease but still has the ability to transfer the pathogens to the healthy host causing the communicable disease is called a Convalescent carrier or the period of convalescence is the period after which symptoms have subsided [15]. Asymptomatic carriers are active carriers that do not display signs or symptoms but have bacteria, virus, fungus, or parasite invaded or present in the body but have not yet caused any symptoms [16]. Maybe our body fights off with them by itself, and we never know it was there, or we may get symptoms after the asymptomatic phase. Pathogens such as herpes simplex virus, hepatitis virus B, Human immune deficiency virus, are frequently transmitted by asymptomatic carriers. For example, with cytomegalovirus (CMV), CMV infection affects about 1% of newborns, although the majority of infections are asymptomatic [17]. A chronic carrier is anyone who harbors and transmits an infectious agent for a prolonged period of time. The bacilli that cause typhoid can live for more than a year without causing any symptoms or illness. A pathogen may harbor more than one reservoir. Humans are also susceptible to infectious agents that have animal reservoirs, a condition well defined by the term 'zoonoses.' Thus, any infectious disease that is naturally trans-missible from vertebrate species to humans is referred to as zoonosis (plural 'zoonoses') [18]. In zoonotic disease, animals act as a reservoir of human disease and transfer to humans by direct or indirect contact. For instance, eating undercooked meat, eggs, or consuming inadequately washed products that may be contaminated with animal feces, drinking raw, unpasteurized milk may also promote the spread of zoonotic diseases to humans. Zoonoses may be of bacterial, viral, and parasitic origin. Some of the known examples are anthrax (sheep), Ebola, West Nile encephalitis, emerging coronavirus (SARS). In some cases, the infectious agent may not show any symptoms but affects the animals. Animals thus play a very important role to keep infectious agents in themselves [19].

1.3   CLASSIFICATION OF 'MODES OF TRANSMISSION'

Microorganisms vary in size, length of time of their life cycle, or time for which they survive on surface or air. Transmission is critical for the study of epidemiology and any disease, and transmission of pathogen is complex and multifaceted [20]. Furthermore, not only the term 'mode' of transmission but also the 'route' of transmission is important for studying disease biology because patterns of contact differ between different populations and classes of the population depending on a variety of factors such as socioeconomic and cultural characteristics [21]. Poor hygienic conditions at the personal and food level, for example, due to a lack of clean water in a given area, can result in increased transmission of infection through the fecal-oral path, as in the case of Cholera. Differences in the occurrence of these diseases among different groups can also be linked to the disease's transmission routes [22]. It is important to note that the two terms 'mode' and 'route' are often used interchangeably which often confuses the two concepts that are used for evaluating the process by which transmission evolves [23]. In general usage, 'mode' means the origin of anything and 'route' is the destination of that starting point. For example, in common usage, 'mode' is used for means of transport (e.g., bus, car, train, and bicycle) is easily different from, 'route' taken to reach their destinations (e.g., via which country, or via which specific international departure). Concerning pathogenic infections, [24] we can say that the 'mode' refers to the path taken by the pathogen to get from point A to point B, while the 'route' refers to the path taken by the pathogen to get from point A to point B, and includes the starting point [25]. Table 1.1 presents one of the classifications for transmission modes, as well as the potential routes associated with each mode.

TABLE 1.1   Classifications of Modes of Transmission Along-with the Possible Routes Associated with Specific Mode

Mode				Route (Example)
Vertical				Cytoplasmic, transplacental, breastfeeding, during vaginal birth
Horizontal	Sexual			Mainly genital-genital, oro-genital, flower to flower
	Non-Sexual	Direct Contact		Skin-skin, kissing, biting, touching
		Airborne		Respiratory tract-respiratory tract
		Indirect	Environmental	Contaminated food, contaminated water-oral, fecal-oral
			Fomites	Clothing-skin, needles-blood, doorknob-hand
			Vector-Borne	Cutaneous penetration, vector fecal deposition, vector identity

Source: Ref. [23].

1.3.1   *HORIZONTAL V/S VERTICAL TRANSMISSION*

By definition, Horizontal transmission refers to the mode of transmission through which an infectious pathogen spreads diseases within the same generation or population, such as transmission from one person to another by direct contact such as body fluid exchange or bodily excretions such as sputum or blood [26]. Whereas 'vertical transmission' is the transmission of an infectious pathogen from one generation to the next, not through evolution, but through milk or the placenta from mother to infant [27].

The various modes of horizontal transmission by which a pathogen transmits a disease are generally categorized as follows [28]:

1. **Contact:** It is the most common mode of pathogen transmission from an infected to a healthy host. Contact mode is further divided into two types: direct and indirect.

i. **Direct Mode:** By coming into close contact with the body surface of an infected person or animal. The physical transfer of pathogens from an infected or colonized individual or animal is known as direct mode [29].

    a. **Person to Person:** In this method, the pathogen (virus, bacteria, and various germs) is spread through direct contact with another person (infected host) [30]. This can occur when a person sneezes, coughs, touches, and kisses someone who is not infected. This infection also occurs when a body fluid got exchanged from sexual contact. A person can transfer the pathogen even when it is asymptomatic.

    b. **Animal to Person:** In this transmission pathogen is transmitted from infected animals to a healthy person by biting, scratched, and even our pet animals can lead to a communicable disease and, in extreme circumstances, can be fatal [31]. Handling animals' wastes can be dangerous for a person, too, e.g., one can get a toxoplasmosis infection by scooping your cat's litter box.

ii. **Indirect Mode:** It involves inanimate objects that got contaminated by pathogens (fomites) from an infected person or reservoirs such as tabletop, doorknob, or faucet handle). For example, you may be infected by germs that an infected person left behind when you then touch your eyes, mouth, or nose without washing your hand [32].

    a. **Through Insect Bites:** Only a few pathogens are transferred by insects, such as mosquitoes, flies, lice, or ticks from host to host. Vectors are the name for these carriers. Mosquitoes, for example, can transmit malaria parasites, and the West Nile virus and deer ticks can transmit bacteria that cause Lyme disease [33].

    b. **Through Contaminated Food and Water:** A person can be infected by contaminated food and water. It is a very good source of transmission of a disease by a single source. For example, many bacteria are present in food and water, and they are transferred very easily by coming in contact with them. Some organisms can survive on contaminated objects for several periods [33].

c. **Contaminated Objects:** Microorganisms (pathogens) remain alive even on the surface of objects from where they can be found and picked up by the next individual who touches the surface of the infected objects. Transmission through this indirect mode can be eliminated by cleaning patient equipment, proper handling, and storage of food with hygienic conditions and storage and dosage of medication should be drawn carefully from multi-dose medication vials [32–34].

2. **Droplet Transmission/Airborne:** The COVID-19 pandemic, which was first reported in Wuhan, China, in December of 2019, is the most recent example of Droplet transmission [35]. COVID-19 is caused by a virus known as SARS-CoV-2 (severe acute respiratory syndrome corona virus-2). Many researchers believe COVID-19 is spread from person to person through aerosols, droplets, and spores [36]. Infected people can spread infectious agents by talking, breathing, coughing, or sneezing. Virus particles are known to be encapsulated in globs of mucus, saliva, and water, and the size of globs influences how they behave in the setting. Bigger particles fall faster, which then evaporates so that they form Droplets. Smaller particle evaporates faster in the form of aerosols, and linger in the air and drifts farther away than the droplets do. Respiratory particles are further distinguished to be droplets or aerosols. The molecules having a size of less than 5 µm are known as aerosols (Droplet nuclei), and more than 5 µm are known as Droplets [36]. Most of the researchers believed that the severity of infection is caused by droplets, not by Aerosols. There are numerous arguments on the average size of droplets and aerosols [36]. Basically, small aerosols have the potential to be inhaled deep into the lung, which causes the infection in alveolar tissues of the lower respiratory tract, as compared to the larger particles of aerosols that get trapped in upper airways. For easy apprehension aerosols are those particles formed by the suspension of solid and liquid particles in the air [37]. However, in recent research, evidence has been provided to deny the former research hypothesis and contemplated that aerosols play a critical role in the transmission of disease. Figure 1.3 is a diagrammatic representation of airborne and droplet infections. There is still

a controversy on modes of transmission that how SARS-CoV-2 is transmitted, including WHO; hence remains an unresolved dichotomy [5, 37].

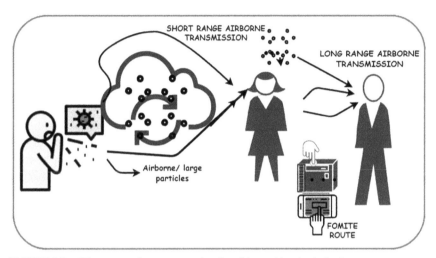

FIGURE 1.3   Diagrammatic representation for airborne/droplet infections.

Direct transmission by an infected person to another is the primary source of aerosols and droplets, aerosols are generated by the waste of medical procedure, wasted tap water, and toilet flushes which is contaminated with infectious agents [37]. Apart from fomites, direct contact transmission caused by Aerosols generating medical procedures (AGMPs) is also a harmful mode of transmission. When an infected person sneezes, speaks, or coughs and comes into direct contact with mucous and conjunctiva, droplets are produced. To infect, a disease could occur by poor hand sanitation or by not following the common disease-controlling courtesy [38, 39]. For example, SARS-CoV-2 is also considered as a case of aerosols transmission or transmission by respiratory droplets [39]. Many pieces of research regarding the mode of transmission proved that SARS mainly transmitted through Aerosols so-called airborne transmission even in the absence of AGMPs, many experimental studies support that by sneezing and coughing, small particles are released which can be suspended in the air and can be traveled for 27 feet, in hospitals scientists collected the sample and found that air sample had aerosols and causative agents of SARS-CoV-2 which shows that COVID-19 is caused by aerosols [40]. Another example of aerosol's transmission is MERS (Middle East respiratory syndrome), many research articles, or guidance shows that MERS is also mainly transmitted through large droplets or aerosols [41]. Many experimental reports suggested that SARS-CoV-2 is mainly transmitted by aerosols (so-called airborne transmission) even in the absence of AGMPs [42].

1.4   EVOLUTION OF TRANSMISSION MODES

Evolutionary changes are not a pre-requisite for the emergence of disease in the case of viruses with a wide host range as found in the case of canine distemper virus where the virus has the potential to infect more mammals, naturally infecting sea mammals, *Mustela nigripes*, lions, and other hosts, and its emergence within these species appears to be limited primarily by contact [43]. However, in other cases, emergence often requires the evolution of viruses to infect and transmit within a new host. Genetic variations in viruses play a very critical role in evolution, as in most of the viruses which are not genetically evolved and transferred to a new host, replicate poorly, and may not have the potential to transmit in cross-species, from here we can say if a virus has good genetic variation, then it has more chances to be transmitted into the new host. This shows that all evolving viruses can cross/switch the host within species. Most of the RNA viruses have a lack of proofreading mechanism, error-susceptible replication, fast replication, short life cycle, and large population [44]. Many DNA viruses, on the other hand, are less variable and are associated with host-virus co-speciation [45]. Some retroviruses, such as the SFV (spumavirus), have nucleotide substitution rates that are much lower than those seen in other RNA viruses [46]. There is a lot of evidence that RNA viruses will co-evolve with their hosts (arenaviruses and hantaviruses) [44]. Recombination is also important for many viruses because it allows them to acquire several genetic changes in a single-phase and can modulate genetic information to create beneficial genotypes and/or suppress deleterious mutations. The pandemics of 1957 H2N2 and 1968 H3N2 influenza A are two examples of reassortment in disease emergence, where segments of the new avian genome were imported to descended H1N1 viruses and formed a hybrid virus causing pandemic [47]. Fujian H3N2 influenza strain is the result of interclade reassortment [47]. Further, the potential for recombination also varies in RNA and DNA viruses. Apart from segmental reassort-ment, negative-stranded RNA viruses rarely exhibit recombination while retroviruses such as HIV possess high rates of recombination. SARS-CoV appears to have arisen from the recombinant of a bat CoV and another virus before infecting human and carnivorous hosts. Mostly deleterious mutation arising out of reassortments or recombination may disrupt the optimal protein structure and functional gene [45]. The Influenza virus, for example, has replication protein (PB1, PB2, and PA) that acts as a complex and changes the combination by reassorting genomic segments to reduce replication rate and necessitates adapted protein from various sources. In host switching, recombination is critical for host evolution. The H3N2 influenza virus, for example, has avian virus HA and PB1 gene fragments [48].

1.5 POSSIBLE INTERVENTIONS RELATED TO TRANSMISSION

Intervention refers to the hindrance to the transmission of a pathogen into the host. These approaches can be split into two groups and are an effec-tive way to break the infection chain [49]:

1. **Preventive Interventions:** These are those which prevent the persons from developing the disease and eliminate the new cases too; and
2. **Therapeutic Interventions:** These are those which help in curing an already occurred disease by use of drugs, to mitigate or stop the effect of disease [50]. Some of the possible interventions are described below:

1.5.1 PREVENTIVE INTERVENTIONS

1. **Vaccine:** These are typically used to treat a particular disease by delivering several doses or a single dose to a person prior to contact with an infectious pathogen. Vaccines assist in the treat-ment or prevention of a pandemic or harmful infection [49]. Being a cost-effective intervention often assists in the production of long-term immune defenses or memory cells against the virus/ pathogen [51]. The vaccination is given to someone who has never

been exposed to pathogens naturally or to very young children to see how vaccines affect them.

2. **Nutritional Interventions:** Human health and pandemic can be determined by food and nutrition. Low-income countries or middle-income countries may be affected by nutrition and food scarcity which is responsible for any pandemic or infectious disease [52]. A country can overcome such deficiencies by using more micronutrients or by a change in agricultural practices.

3. **Maternal and Neonatal Interventions:** A mother's well-being and health matters when a mother laboring a child. Preventive interventions of pregnant women before or during pregnancy include treatment of disease and planning of family, infectious diseases such as malaria, typhoid, and syphilis, healthy nutrition [51]. By good antenatal care and the health of pregnant women, we can eliminate the interventions of transmission.

1.5.2   THERAPEUTIC INTERVENTIONS

- **Treatment of Disease:** For trials, we simply focus on drug mechanisms to know how the drug molecule will behave with the patient and its negative impacts on the infected or non-infected person [53]. Drugs are designed to eliminate the effect of disease and to inhibit the replication of a pathogen in host cells which help in eliminating or minimizing mortality and morbidity. For example, diseases such as leprosy and tuberculosis, if a person infected by these kinds of disease is not getting proper drug or treatment which can increase the chances of infection to the healthy host so by providing treatment on time, we can eliminate the spreading of disease [52].

1.6   HOST SWITCHING/CROSS-SPECIES TRANSMISSION

A Host Switch occurs when a pathogens or infectious agent's host specificity changes as a result of evolution. For example, HIV first infected the non-human primates in West Africa, and after replication and evolutionary change, it skipped to humans and caused AIDS. Newly emerging viruses and parasites are a threat to public health [53]. Some wildlife viruses cause emerging diseases that are highly harmful to people's health, such as SARS, which is transmitted by bats, and many others, such as Ebola and influenza in humans. Human diseases occur when an established virus switch their hosts to humans and transferred to the human population [52]. Switching hosts within different animals leads to the emergence of epizootic disease, a common example of an epizootic virus is the H5N1 avian virus, influenza A. H5N1 virus thrives in a potentially active host and switching to a new human host caused pandemic and posed a threat for humans [54].

There are three stages of host switching that may lead to the emergence of a disease:

- **Stage 1:** In this stage, there is no more transmission after initial infection to the new host;
- **Stage 2:** At this stage, the virus gets spread and cause local transmission with new host generations before pandemic; and
- **Stage 3:** At this stage host to host transmission take place in a new host population.

Epizootic and enzootic diseases are the main sources of human viral infections. We know only a few viruses that are infecting even wild or domesticated animals [55]. Such unrecognized viruses have been identified by the emergence of many viruses such as the human immune deficiency virus, Nipah virus, SARS-CoV, Ebola, and Hantavirus, involving cross-species host switching that was unknown before their emergence into humans [54]. Some enzootic viruses have the potential of switching the host in between cross-species. HIV or AIDS, which first appeared in ancient primates and then spread to humans, is an example of host switching [56]. A recent example of host switching is CoV causing COVID-19.

1.6.1   *ENVIRONMENTAL AND DEMOGRAPHIC BARRIERS*

The most critical step in moving a pathogen from one host to another is cross-host exposure. Since there is very little interaction between the host and the virus, and host switching is reduced [53]. For example, human immune deficiency viruses 1 and 2 are transmitted to humans'

multiple times but failed to switch the host because of some environmental or demographic barriers. So demographic is a major barrier to
switch the host [57]. In some cases, the virus transfer, which is in the pre-
conditioning phase between donor and recipient is affected by ecological,
geographical, and behavioral factors that may affect the host donor and
recipient. Factors affecting the geographical distribution of host species
such as wildlife trade and introduction of host species, human-induced
changes may promote viral emergencies such as changes in social and
demographic factors such as human travel, population expansion, human
behavior such as sexual practices, and use of drugs [58]. Further, the
patterns of host contact and density also controls the emergence of disease.
Any transferred virus's onward dissemination and disease potential are
influenced by the density of the recipient population. The spread of the
SARS pathogen has been linked to human trade and travel patterns [57].

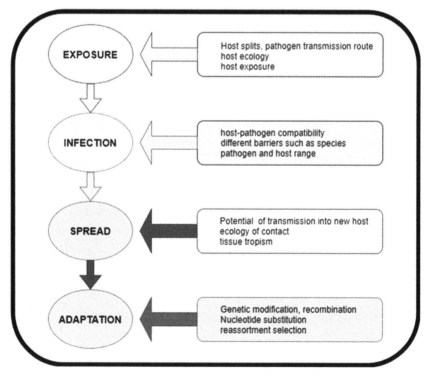

FIGURE 1.4   Sequence of steps involved in host switching process during emergence of
infection.

Another important factor determining the emergence of disease is the role of intermediate and amplifier hosts. They bring the animal virus in close contact with alternative hosts which would otherwise have little contact with the virus, e.g., for SARS-CoV, the virus is believed to have begun in bats and then spread to humans, civet cats, and other carnivores. Although the precise route of transmission is unknown, it is likely that infection of domesticated animals resulted in increased human exposure [58, 59]. Figure 1.4 represents the general outline of steps involved in viral host switching.

1.6.1.1   HOST AS A BARRIER

It is crucial to note that for infection into a new host, in certain cases, the host acts as a virus transfer shield. For example, if a virus wishes to invade a new host, it must infect the new host's unique cells efficiently, and this process can be manipulated at various levels, including genome replication, fusion with the host, entry, signaling inside the cell, gene expression, and receptor binding [56]. Generating and shedding of the virus are very specific to the host. In response to multiple host barriers, viruses are also prone to changes in their self, to cross the host barriers. Some of the important impediments to infection include cytokine and interferon-mediated antiviral responses [56].

Host switching also takes place in between closely related or distantly related genes [60]. For example, HIV is transferred from chimpanzee to human because they are more related host that causes epidemics or spillover infection and another example is CoVs which are transmitted in between closely related animals such as from bats to human as well as to civets and other carnivores [60]. The entry of a virus into a new host is a key step in deciding the virus's host specificity, and a shift in receptor binding will prevent the virus from communicating with the host [56]. SARS virus, for example, circulates in a bat reservoir but changes hosts when it comes into contact with humans or carnivorous angiotensin-converting enzyme two receptors [56].

1.7   TRANSMISSION CYCLE OF SARS-COV AND SARS-COV-2

Severe acute respiratory syndrome (SARS) is a dreadful disease caused by the CoVs, namely SARS-CoV and is characterized by some specific symptoms such as headache, extreme fever, and many severe symptoms of the respiratory tract such as short breath, pneumonia, and dry cough [61]. The pathogenic CoVs can infect humans and livestock such as mice, bats, and birds at various sites such as the gastrointestinal, respiratory, hepatic, and central nervous systems. Previously SARS-CoV belongs to the group of 2b (gene) CoV and now it is a member of lineage B of genus β coronavirus belonging to the family of Coronaviridae and subfamily Coronavirinae [61]. The genome of SARS-CoV shares similarity to other coronaviruses, but have some specific unique genes including 3b, ORF3a, ORF7a, ORFa, 7b, ORF 8a, 8b, and 9b. Importantly, SARS-CoV interacts with the ACE-2 receptor of the host cell to infect bronchial epithelial cells and type II pneumocytes in the host. A recent spillover or outbreak of CoV strain (COVID-19) in China, in 2019, has gained attention all over the world, as continuous evolution and transformation led to the emergence of pandemic all over the world [62].

Transfer of animal diseases which normally occurs in animals and infects the human are called Zoonotic diseases. For a century, large populations have been affected by several Zoonotic diseases, among which lies SARS, MERS, and COVID-19. However, with time passing, they have evolved and mutated, in several perspectives. Pieces of evidence indicate that transmission of CoV also occurs via 'Zoonotic Spillover.' The latter depends upon both characteristics of CoV and susceptibility of host [63].

In the case of SARS, the first case of SARS-CoV transmission was reported in Guangdong province, southern China, from vertebrate to humans, from zoonotic reservoirs, such as raccoon dogs, civets, and bats. SARS-CoV is transmitted to humans from market civets, with bats as the possible reservoirs. Further, SARS-CoV can infect macaques and ferrets and cause disease, while remaining asymptomatic in cats. In recent research, there is an 80% gene similarity between SARS-CoV and SARS-CoV-2 and one study showed that 96% of genes are similar in *Rhinophus affinis* showing that bat as virus source. Scientists are failed to found the host of CoV-2 yet, some researchers stated that hosts for SARS-CoV-2 are snakes, minks, and many other animals [61, 64].

As per the current evidence, the natural host of CoV is the bat, whereby CoV has evolved to bind the receptor of the host through its surface's glycoproteins. Variations in these surface glycoproteins, allow the virus to bind to a variety of hosts. However, another important feature of these CoVs is that CoVs may avoid the receptor-mediated entry into the host cell rather they may employ a different non-receptor mediated entry into the host cell [63]. The latter is mediated by, a kind of receptor switching mechanism which involves spike modularity and recombination. The presence of potent fusogenic potential in Spike protein may facilitate the infection via cell-to-cell spread mechanism. For instance, The JHM (MHV-JHM) murine CoVs strain, which codes for a highly fusogenic spike protein, requires infection to spread through cell-to-cell transmission rather than receptor-dependent entry [65].

1.7.1   TRANSMISSION IN ANIMALS

As per the reports of 2005, the CoVs associated with humans SARS-CoV, were initially named as SARS-like CoVs were found in horseshoe bats (genus *Rhinolophus).* According to this research, it was understood that bat is a natural host or reservoir for this CoV, while civets worked as an intermediate reservoir of CoVs [64]. Many researchers stated that bats play a very critical role as hosts, and due to recombination between bats SARS-CoV existing either in the same or another bat caves, there is a high rate of emergence of new SARS-CoV. One of the possible explanations is that SARS-CoV direct progenitor was produced by recombination events in bats, and thereafter, the infection passed on to farm civets and other mammals, by fecal-oral transmission. The civets harboring the virus were transported to the Guangdong market, where market civets were infected, and then further mutations were produced in the virus to infect humans [66].

1.7.2   TRANSMISSION TO HUMAN FROM ANIMALS

As mentioned above the SARS-CoV-2 virus originated from bats in Wuhan, China. Current research confirms that more than 95% of genes are similar in SARS-CoV, and SARS-CoV-2, indicates that the most probable reservoir of CoV-2 is also bat [67]. One more remarkable phenomenon observed was that SARS-CoV recovered in human at the time of epidemic had efficient hACE2/cACE2 and *in vitro* adapted strain of civet also have receptor recognition of hACE2 and cACE2 [60] from where we can say that human/civet ACE2 recognition is a key factor for transmission from animals to humans, and animal reservoir supports the continual persistence for CoVs [68]. The spillover event thus takes place from bats to civets and later on to humans. Further, a spillover is also comprised of certain consecutive events in order to establish the infection in humans. Transmission from animal to humans is governed by certain factors and divided into three stages:

- **Stage 1:** It is controlled by the prevalence of the virus, dispersal from the animal host, survival, growth, and dissemination beyond the animal host, and is expressed in terms of pathogen pressure on the human host, shown by the amount of virus interacting with humans at a given point in time.
- **Stage 2:** Defined by viral exposure, route of entry, and a dose of virus, the actions of vectors and the human host are both variables that affect the outcome [67].
- **Stage 3:** Defined by Possibility (in combination with stage 2 also) and severity of infections-governed by the genetics factors, physiological, and immunological status of the human host.

Each stage corresponds to a barrier that a virus must cross (during spillover) in order to infect the next host [66].

1.7.3 *TRANSMISSION AMONG HUMANS*

Both SARS-CoV and SARS-CoV-2 have zoonotic origins having fast human-to-human transmission. Like cold and flu, SARS is also transmitted from an infected person by coughing and sneezing through droplet transmission or via stool. As per the recent reports SARS-CoV-2, have been detected in the stool of infected person which gives the evidence of gastrointestinal infection and it also suggests that this virus is transmitted to human by contaminated food or feces-oral route. Many researchers believed that this virus is transmitted to the person upon prolonged exposure to viruses; it may be because of droplets by the infected person like sneezing, and coughing [61].

Till now, a sustained human-to-human transmission has been reported for SARS-CoV-2 with SARS-CoV-2 infected person as the major source of infection and respiratory droplets and aerial droplets (airborne infection), close contact as the main route of transmission.

1.8 TRANSMISSION STATUS OF COVID-19

There are some shreds of evidence that provide some additional information about the transmission of COVID-19. COVID-19 is caused by the SARS-CoV-2, as has been mentioned above that it spreads via respiratory droplets and close contact. However, these types of modes of transmission do not provide concrete evidence regarding the transmission of this disease. But there are many experiments held at Wuhan to understand the transmission of SARS-CoV-2 like a patient who is infected and asymptomatic can spread the virus without any symptoms so from here we can say direct contact or indirect contact (such as [69] common vehicle like surfaces) might be responsible for spillover of this SARS-CoV-2 [70]. Many theories have evidence that SARS-CoV-2 spread in Wuhan from seafood and bats. As far today, there is no fully elucidated mechanism about transmission and natural host of the virus, and the intermediate host who infects the human is still in question. However, the disease is known to be transmitted from human to human and led to the spread of pandemic COVID-19 through respiratory droplets such as sneezing, coughing [70, 71]. Apart from the initially recognized modes of transmission, which were mainly due to airborne or droplet mediated, the recent investigations have reported some other possible modes of transmission. A few of them are described in subsections.

1.8.1 TRANSMISSION BY THE ENVIRONMENT, STOOL, AND SEWAGE

SARS-CoV-2 RNA has been detected in stools of positive patients of COVID-19 whether the infections were symptomatic or asymptomatic, which indicates that transmission can occur via raw sewage network. A person who has diarrhea possesses a higher possibility to transfer the virus because the virus is more stable in higher pH, nevertheless, the virus is detected in many patients who do not have diarrhea, and around 97% of patients had the virus present in their stool [72].

1.8.2 AIR POLLUTION

We can relate the relationship between COVID-19 and air pollution, first, we can see from the lockdown procedure that when the lockdown was implemented all over the cities, the quality of air changed, which gives positive aspects related to COVID-19. Air pollution has been linked to COVID-19 through the impact of poor AQI on human health and more susceptibility to the virus [38, 73]. As there is no substantial evidence regarding the air pollutants such as PM and NO_2 acting as a vector despite many scientists support this idea but future investigations are still required to conclude. As pollutants cause inflammation, respiratory problems, cellular damage, and eliminate the response to pathogens hence many pollutants such as PM10 and PM2.5 may serve as a causative factor for pneumonia and chronic pulmonary disease [74].

1.8.3   OCULAR TRANSMISSION

SARS-CoV-2 has also been detected in the ocular surface of infected patients and conjugative secretions. Nevertheless, in a recent study, it was found that even all those who do not have conjugative secretions can spread the disease. Although the study report found evidence of virus to be present in tears, it shows that ocular involvement in SARS-CoV-2 might also apply to COVID-19 and suggested that direct involvement of ocular transmission caused infection through nasolacrimal ducts and hematogenous infection of lacrimal glands [75, 76].

1.8.4   AIRBORNE/AEROSOL

Airborne is still an unknown source of transmission, there are numerous studies and debates on whether a virus can be transmitted through aerosols. It was found that small particles which are less than 5 μm tend to split in air and chances to transfer by air while larger particle gets to settle down by the force of gravity within 1–2 m radius. Many diseases are caused by airborne like tuberculosis, influenza, SARS, varicella, and measles. Several studies approved that SARS-CoV-2 particles are found in the air and have the potential to transfer by air. In a study, researchers found that 0.98–8.69 copies/L of the virus are found in the room of positive patients in the air [73].

1.8.5   SARS-COV-2 IN SOIL

Many researchers provided the evidence that wastewater and sludge contain RNA of SARS-CoV-2 and the environment also contains SARS, several observations have been reported, including various papers that are hypothetically published but practically do not have data on soil transmission. Now it is a very important question that is soil responsible for the transmission of COVID-19 or not??? To answer this question, we have some aspects like wastewater that may be directly applied in soil, this could be wastewater or treated wastewater which might be used for irrigation, but this wastewater does not have a virus [77, 78].

1.9 CONCLUDING REMARKS

Modes of transmission form an integral component for the global spread of pandemic diseases such as SARS and COVID-19. In the case of SARS-CoV and SARS-CoV-2, the suspected zoonotic origin and subsequent spillover on other animals and humans requires further study, as animal to host switching is a major source of newly emerging infectious diseases. In light of current evidence available for the emergence of the COVID-19 pandemic, the modes of transmission lie under a vast range, contributing towards its global spread at a very fast pace. In such cases, the interface between the virus and the host (humans) is of paramount importance. Future research investigations should focus on molecular and cellular level targets as well as routes of infection to design preventive/therapeutic strategies for breaking the chains of infection.

ACKNOWLEDGMENTS

Dr. Neha Gupta acknowledges the Department of Science and Technology (DST), Government of India, for providing support through the INSPIRE FACULTY Fellowship Award [DST/04/2017/002738].

KEYWORDS

- **aerosols generating medical procedures**
- **coronavirus transmission**
- **COVID-19**
- **cross-species transmission**

- **middle east respiratory syndrome**
- **modes of transmission**
- **pandemic**
- **severe acute respiratory syndrome**

REFERENCES

1. Thomas, Y., et al., (2008). Survival of influenza virus on banknotes. *Applied and Environmental Microbiology, 74*(10), 3002–3007.
2. WHO/UNICEF Joint Water Supply, (2005). *Sanitation Monitoring Program, World Health Organization, WHO/UNICEF Joint Monitoring Program for Water Supply, Sanitation, and UNICEF.* Water for life: making it happen. World Health Organization.
3. Brankston, G., Gitterman, L., Hirji, Z., Lemieux, C., & Gardam, M., (2007). Transmission of influenza A in human beings. *The Lancet Infectious Diseases, 7*(4), 257–265.
4. Marineli, F., Tsoucalas, G., Karamanou, M., & Androutsos, G., (2013). Mary Mallon (1869–1938) and the history of typhoid fever. *Annals of Gastroenterology: Quarterly Publication of the Hellenic Society of Gastroenterology, 26*(2), 132.
5. Cobb, S., Miller, M., & Wald, N., (1959). On the estimation of the incubation period in malignant disease: The brief exposure case, leukemia. *Journal of Chronic Diseases, 9*(4), 385–393.
6. Brooks, J., (1996). The sad and tragic life of Typhoid Mary. *CMAJ: Canadian Medical Association Journal, 154*(6), 915.
7. Ayres, C. N. F. J., (2016). Identification of Zika virus vectors and implications for control. *The Lancet Infectious Diseases, 16*(3), 278, 279.
8. Leavitt, J. W., (1996). *Typhoid Mary: Captive to the Public's Health.* Beacon Press.
9. Remington, P. L., Hall, W. N., Davis, I. H., Herald, A., & Gunn, R. A., (1985). Airborne transmission of measles in a physician's office. *JAMA, 253*(11), 1574–1577.
10. Poutanen, S. M., & Simor, A. E., (2004). Clostridium difficile-associated diarrhea in adults. *CMAJ, 171*(1), 51–58.
11. McDonald, L. C., et al., (2005). An epidemic, toxin gene-variant strain of clostridium difficile. *New England Journal of Medicine, 353*(23), 2433–2441.
12. White, D. J., et al., (1991). The geographic spread and temporal increase of the Lyme disease epidemic. *JAMA, 266*(9), 1230–1236.
13. Parker, E. R., Dunham, W. B., & Macneal, W. J., (1944). Resistance of the Melbourne strain of influenza virus to desiccation. *Journal of Laboratory and Clinical Medicine, 29*(1).
14. Soper, G. A., (1939). The curious career of typhoid Mary. *Bull. N Y Acad. Med., 15*, 698–712.

15. Furuya-Kanamori, L., Cox, M., Milinovich, G. J., Magalhaes, R. J. S., Mackay, I. M., & Yakob, L., (2016). Heterogeneous and dynamic prevalence of asymptomatic influenza virus infections. *Emerging Infectious Diseases, 22*(6), 1052.

16. Heffernan, R. T., Pambo, B., Hatchett, R. J., Leman, P. A., Swanepoel, R., & Ryder, R. W., (2005). Low seroprevalence of IgG antibodies to Ebola virus in an epidemic zone: Ogooue-Ivindo region, Northeastern Gabon, 1997. *The Journal of Infectious Diseases, 191*(6), 964–968.

17. Shenoy, E. S., Paras, M. L., Noubary, F., Walensky, R. P., & Hooper, D. C., (2014). Natural history of colonization with methicillin-resistant *Staphylococcus aureus* (MRSA) and vancomycin-resistant enterococcus (VRE): A systematic review. *BMC Infectious Diseases, 14*(1), 1–13.

18. Finkbeiner, A. K., (1996). Quite contrary. *Sciences, 36*(5), 38.

19. Usman, M., Farooq, M., & Hanna, K., (2020). *Existence of SARS-CoV-2 in Wastewater: Implications for its Environmental Transmission in Developing Communities.* ACS Publications.

20. DeMuri, G. P., & Wald, E. R., (2014). The group A streptococcal carrier state reviewed: Still an enigma. *Journal of the Pediatric Infectious Diseases Society, 3*(4), 336–342.

21. Beaglehole, R., Bonita, R., & KjellstrÃm, T., (1993). *Basic Epidemiology.* World Health Organization Geneva.

22. Celentano, D. D., & Szklo, M., (2015). In: Leon, G., (ed.), *American Journal of Epidemiology* (Vol. 182, No. 10, pp. 823–825).

23. Antonovics, J., et al., (2017). The evolution of transmission mode. *Philosophical Transactions of the Royal Society B: Biological Sciences, 372*(1719), 20160083.

24. Park, K., (2005). Park's textbook of preventive and social medicine. *Preventive Medicine in Obstet, Pediatrics and Geriatrics.*

25. Kuno, G., & Chang, G. J. J., (2005). Biological transmission of arboviruses: Reexamination of and new insights into components, mechanisms, and unique traits as well as their evolutionary trends. *Clinical Microbiology Reviews, 18*(4), 608–637.

26. Twiddy, S. S., Pybus, O. G., & Holmes, E. C. J. I., (2003). *Genetics, and Evolution: Comparative Population Dynamics of Mosquito-Borne Flaviviruses, 3*(2), 87–95.

27. Weaver, S. C., & Barrett, A. D. T., (2004). Transmission cycles, host range, evolution and emergence of arboviral disease. *Nature Reviews Microbiology, 2*(10), 789–801.

28. Weber, F., Kochs, G., & Haller, O., (2004). Inverse interference: How viruses fight the interferon system? *Viral Immunology, 17*(4), 498–515.

29. Weiss, R. J. X. T., (2003). *Cross-Species Infections*, 47–71.

30. Van, S. J. M., & Hochberg, N. S., (2017). Principles of infectious diseases: Transmission, diagnosis, prevention, and control. *International Encyclopedia of Public Health, 22.*

31. CDC: Centers for Disease and Prevention, (2006). In: Atlanta, G. A., (ed.), *Principles of Epidemiology in Public Health Practice: An Introduction to Applied Epidemiology and Biostatistics.* US Dept. of Health and Human Services, Centers for Disease.

32. Getz, W. M., & Pickering, J., (1983). Epidemic models: Thresholds and population regulation. *The American Naturalist, 121*(6), 892–898.

33. Antonovics, J., (2017). Transmission dynamics: Critical questions and challenges. *Philosophical Transactions of the Royal Society B: Biological Sciences, 372*(1719), 20160087.

34. De Vienne, D., Refrégier, G., López-Villavicencio, M., Tellier, A., Hood, M., & Giraud, T. J. N. P., (2013). *Cospeciation vs Host-Shift Speciation: Methods for Testing, Evidence from Natural Associations and Relation to Coevolution, 198*(2), 347–385.

35. Thom, K., Morrison, C., Lewis, J. C. M., & Simmonds, P., (2003). Distribution of TT virus (TTV), TTV-like minivirus, and related viruses in humans and nonhuman primates. *Virology, 306*(2), 324–333.

36. Belingheri, M., Paladino, M. E., & Riva, M. A., (2020). Beyond the assistance: Additional exposure situations to COVID-19 for healthcare workers. *Journal of Hospital Infection, 105*(2), 353.

37. Wang, M., et al., (2020). *Temperature Significant Change COVID-19 Transmission in 429 Cities*. Medrxiv.

38. Wang, J., & Du, G., (2020). COVID-19 may transmit through aerosol. *Irish Journal of Medical Science (1971), 1*, 2.

39. Wenzel, R. P., & Edmond, M. B., (2003). Managing SARS amidst uncertainty. *New England Journal of Medicine, 348*(20), 1947, 1948.

40. Tellier, R., Li, Y., Cowling, B. J., & Tang, J. W., (2019). Recognition of aerosol transmission of infectious agents: A commentary. *BMC Infectious Diseases, 19*(1), 1–9.

41. Thomas, R. J., (2013). Particle size and pathogenicity in the respiratory tract. *Virulence, 4*(8), 847–858.

42. Tellier, R., (2009). Aerosol transmission of influenza A virus: A review of new studies. *Journal of the Royal Society Interface, 6*(6), S783–S790.

43. Surveillances, V., (2020). The epidemiological characteristics of an outbreak of 2019 novel coronavirus diseases (COVID-19)-China, 2020. *China CDC Weekly, 2*(8), 113–122.

44. Cleaveland, S., Laurenson, M. K., & Taylor, L. H., (2001). Diseases of humans and their domestic mammals: Pathogen characteristics, host range and the risk of emergence. *Philosophical Transactions of the Royal Society of London. Series B: Biological Sciences, 356*(1411), 991–999.

45. Domingo, E., & Holland, J. J., (1997). RNA virus mutations and fitness for survival. *Annual Review of Microbiology, 51*(1), 151–178.

46. Holmes, E. C., (2004). The phylogeography of human viruses. *Molecular Ecology, 13*(4), 745–756.

47. Switzer, W. M., et al., (2005). Ancient co-speciation of simian foamy viruses and primates. *Nature, 434*(7031), 376–380.

48. Holmes, E. C., et al., (2005). Whole-genome analysis of human influenza A virus reveals multiple persistent lineages and reassortment among recent H3N2 viruses. *PLoS Biol., 3*(9), e300.

49. Lindstrom, S. E., Cox, N. J., & Klimov, A., (2004). Genetic analysis of human H2N2 and early H3N2 influenza viruses, 1957–1972: Evidence for genetic divergence and multiple reassortment events. *Virology, 328*(1), 101–119.

50. Brown, C. A., & Lilford, R. J., (2006). The stepped wedge trial design: A systematic review. *BMC Medical Research Methodology, 6*(1), 1–9.

51. Craig, P., Dieppe, P., Macintyre, S., Michie, S., Nazareth, I., & Petticrew, M., (2008). Developing and evaluating complex interventions: The new medical research council guidance. *BMJ, 337.*

52. Jaffar, S., et al., (2009). Rates of virological failure in patients treated in a home-based versus a facility-based HIV-care model in jinja, southeast Uganda: A cluster-randomized equivalence trial. *The Lancet, 374*(9707), 2080–2089.

53. Peeling, R. W., Smith, P. G., & Bossuyt, P. M. J. N. R. M., (2006). *A Guide for Diagnostic Evaluations, 4*(9), S2–S6.

54. Wu, Z., & McGoogan, J. M., (2020). Characteristics of and important lessons from the coronavirus disease 2019 (COVID-19) outbreak in China: Summary of a report of 72 314 cases from the Chinese center for disease control and prevention. *JAMA, 323*(13), 1239–1242.

55. Ferguson, N. M., et al., (2005). *Strategies for Containing an Emerging Influenza Pandemic in Southeast Asia, 437*(7056), 209–214.

56. Murray, C. J. L., Lopez, A. D., Chin, B., Feehan, D., & Hill, K. H., (2006). Estimation of potential global pandemic influenza mortality on the basis of vital registry data from the 1918–1920 pandemic: A quantitative analysis. *The Lancet, 368*(9554), 2211–2218.

57. Woolhouse, M. E. J., Haydon, D. T., & Antia, R., (2005). Emerging pathogens: The epidemiology and evolution of species jumps. *Trends in Ecology & Evolution, 20*(5), 238–244.

58. Wolfe, N. D., Daszak, P., Kilpatrick, A. M., & Burke, D. S., (2005). Bushmeat hunting, deforestation, and prediction of zoonotic disease. *Emerging Infectious Diseases, 11*(12), 1822.

59. Heeney, J. L., Dalgleish, A. G., & Weiss, R. A., (2006). Origins of HIV and the evolution of resistance to AIDS. *Science, 313*(5786), 462–466.

60. Wong, S., Lau, S., Woo, P., & Yuen, K. Y., (2007). Bats as a continuing source of emerging infections in humans. *Reviews in Medical Virology, 17*(2), 67–91.

61. Jin, Y., et al., (2020). *Virology, Epidemiology, Pathogenesis, and Control of COVID-19, 12*(4), 372.

62. Li, W., et al., (2005). Bats are natural reservoirs of SARS-like coronaviruses. *Science, 310*(5748), 676–679.

63. Li, B., et al., (2020). *Discovery of Bat Coronaviruses Through Surveillance and Probe Capture-Based Next-Generation Sequencing* (Vol. 5, No. 1). Msphere.

64. Ji, W., Wang, W., Zhao, X., Zai, J., & Li, X. J. J. M. V., (2020). *Homologous Recombination within the Spike Glycoprotein of the Newly Identified Coronavirus May Boost Cross-Species Transmission from Snake to Human, 92*(4), 433–440.

65. Lau, S. K. P., et al., (2005). Severe acute respiratory syndrome coronavirus-like virus in Chinese horseshoe bats. *Proceedings of the National Academy of Sciences, 102*(39), 14040–14045.

66. Ge, X. Y., et al., (2013). Isolation and characterization of a bat SARS-like coronavirus that uses the ACE2 receptor. *Nature, 503*(7477), 535–538.

67. Rothan, H. A., & Byrareddy, S. N., (2020). The epidemiology and pathogenesis of coronavirus disease (COVID-19) outbreak. *Journal of Autoimmunity, 109*, 102433.

68.   Wan, Y., Shang, J., Graham, R., Baric, R. S., & Li, F. J., (2020). *Receptor Recognition by the Novel Coronavirus from Wuhan: An Analysis Based on Decade-Long Structural Studies of SARS Coronavirus, 94*(7).

69.   Bibby, K., & Peccia, J., (2013). Identification of viral pathogen diversity in sewage sludge by metagenome analysis. *Environmental Science and Technology, 47*(4), 1945–1951.

70.   Randazzo, W., Truchado, P., Cuevas-Ferrando, E., Simón, P., Allende, A., & Sánchez, G. J. W. R., (2020). *SARS-CoV-2 RNA in Wastewater Anticipated COVID-19 Occurrence in a Low Prevalence Area, 181*, 115942.

71.   Trajano, G. D. S. D., Ives, K., Fesselet, J. F., Ebdon, J., & Taylor, H. J. W., (2019). *Assessment of Recommendation for the Containment and Disinfection of Human Excreta in Cholera Treatment Centers, 11*(2), 188.

72.   Kitajima, M., et al., (2020). *SARS-CoV-2 in Wastewater: State of the Knowledge and Research Needs*, 139076.

73.   Wang, Y., Yuan, Y., Wang, Q., Liu, C., Zhi, Q., & Cao, J. J. S., (2020). *Changes in air Quality Related to the Control of Coronavirus in China: Implications for Traffic and Industrial Emissions, 731*, 139133.

74.   Wu, X., Nethery, R. C., Sabath, B. M., Braun, D., & Dominici, F. J. M., (2020). Air pollution and COVID-19 mortality in the United States: Strengths and limitations of an ecological regression analysis. *Science Advances, 6*(45), eabd4049.

75.   Ho, D., et al., (2020). *COVID-19 and the Ocular Surface: A Review of Transmission and Manifestations, 28*(5), 726–734.

76.   Xie, H. T., et al., (2020). *SARS-CoV-2 in the Ocular Surface of COVID-19 Patients, 7*(1), 1–3.

77.   Bofill-Mas, S., & Rusiñol, M., (2020). Recent trends on methods for the concentration of viruses from water samples. *Current Opinion in Environmental Science and Health, 16*, 7–13.

78.   Everard, M., Johnston, P., Santillo, D., & Staddon, C., (2020). The role of ecosystems in mitigation and management of COVID-19 and other zoonoses. *Environmental Science and Policy, 111*, 7–17.

CHAPTER 2

Understanding the Role of Existing Technology in the Fight Against COVID-19

VINAMRITA SINGH

Netaji Subhas University of Technology (formerly Ambedkar Institute of Advanced Communication Technologies and Research, Government of NCT of Delhi), Delhi, India, E-mails: drvinamrita@aiactr.ac.in; vinamritasingh.phy@gmail.com

ABSTRACT

The emergence of SARS-CoV-2 has shaken the whole world. The existing medical facilities proved to be inadequate for handling the abrupt burden of the infected population. The spread of COVID-19 was so fast that the lack of information, testing tools, protective equipment, treatment protocols, etc., posed a challenge to combat the disease. The fight against the pandemic situation necessitates thorough investigation from the fundamental point of view. The study of the novel virus requires the use of advanced technologies, which would assist in the diagnosis, management, sanitization, and overcoming the shortage of essential supplies. In this context, the present chapter discusses the working principle of various technologies that are helpful with respect to COVID-19. Details about the mechanism of thermal screening, spectroscopy techniques, imaging techniques, biosensors, sanitization, environmental monitoring, and 3D printing technology are discussed. This provides a background about the different devices, which can be used for combating the present and future outbreaks.

2.1 INTRODUCTION

The sudden and unforeseen upsurge of an unknown virus has taken aback the whole world. With what started as an epidemic in Wuhan, China, in December 2019, it hardly took months for the constricted problem to become a global pandemic situation. A situation so grave that all the countries are unitedly fighting against the deadly virus known as severe acute respiratory syndrome (SARS) coronavirus 2 (SARS-CoV-2). An increasing amount of research work is being done all over the world to understand the nature, characteristics, and implications of coronavirus disease (COVID-19) caused by SARS-CoV-2. The research focuses not only on how to treat the patients suffering from COVID-19, but also on the aspects related to its diagnosis, spread, and prevention. Therefore, an amalgamation of different technologies is needed in order to mitigate the consequences of COVID-19, and for better preparedness during future recurrences.

Considering the importance of integrating different technologies to fight against virus attacks in a fast-paced manner, this chapter focuses on the available tools which can be utilized in the time of need. The current methods implemented for primary screening up to patient care are discussed from the fundamental point of view. The physics and fundamental concepts that are used to identify, image, and diagnose COVID-19 will be presented. The chapter concentrates on the mechanisms through which many facets of physics like spectroscopy techniques, photonics, and optoelectronics come into play. For example, a technique like Raman spectroscopy is a very fast, cost-effective method to identify biomarkers, and may change the way current testing is being done. It has the potential to increase the sample testing rate by many folds so that infected persons can be identified early. The use of non-contact thermometers spiked within weeks of the outbreak, which is based on the use of infrared (IR) radiations. Further, the role of biosensors in detecting viruses is presented, which gives a future direction for fast mass screening. The virus also spreads in the wastewater, and contaminates the water bodies. In order to contain or destroy it, large scale technology is required such as ultra-violet light-based sanitization, and nanomaterials-based filters. The chapter will also briefly discuss the role of 3D printing and why it is gaining popularity during this critical phase.

2.2   INFRARED (IR) THERMOMETERS

The infrared (IR) thermometers were widely investigated after World War II and spanning through the Cold War. The handheld, affordable devices came into the market in the early 2000s, and have since been used in a large number of applications. However, the recent outbreak of COVID-19 has spiked its popularity due to its contactless measurements. The IR thermometer measures the surface temperature of any object without physical contact between the object and the thermometer. The temperature measurement using IR thermometers is very fast, has good accuracy, can be done from a distance, and helps in avoiding repeated sanitization of the device. This has helped in preliminary screening at public places during the coronavirus outbreak by scrutinizing people who tend to have high body temperature. The use of IR thermometers has been commonly observed at the entrances/exits of public buildings, offices, shopping centers, and airports.

The working principle of an IR thermometer is based on the black body radiations. All bodies with mass emit heat energy in the IR region. The IR thermometer detects the heat energy (IR electromagnetic radiations) emitted by an object and compares it with the IR radiations coming from the surrounding environment. The difference between the two incoming radiations gives a measure of the temperature. According to Stefan-Boltzmann's law, the power radiated by a body is proportional to the fourth power of temperature. The main components of an IR thermometer comprise a lens and a detector. The lens focuses the incoming radiations onto the detector, which converts the radiations into electrical energy. The detector used in the IR thermometer is a thermopile. A thermopile consists of several thermocouples connected in series. The thermopile absorbs the IR radiations, which heats up one set of the thermocouple junctions. The other junctions are maintained at a reference temperature. The difference in the heat energy (or temperature) between the two layers of thermopile generates an electrical signal. An output voltage is generated which is proportional to the temperature difference, and thus gives the required temperature reading. The total process of temperature measurement takes a few seconds, thus making IR thermometers much faster than conventional medical thermometers. The schematic diagram showing the working of an IR thermometer is shown in Figure 2.1(a).

The distance-to-spot ratio (D:S) is a key parameter that determines the range and accuracy of an IR thermometer. For example, a D:S ratio of 4:1 denotes that the temperature can be accurately measured from a distance of 4 inches for a 1-inch circle. The higher the ratio, the farther the distance from which the temperature can be accurately measured for a narrower surface. The size of the object should be at least twice the spot size for accurate temperature measurement. The measurement error of simple IR thermometers is usually ±2°C. The distance of the object from the thermometer should be within the range specified for the particular instrument, otherwise, a lower accuracy will result due to detection of radiations from surrounding matter.

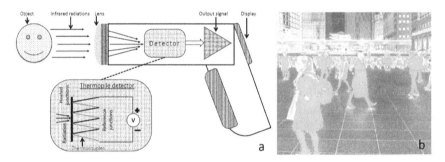

FIGURE 2.1 (a) Simplified diagram of the working of an infrared thermometer (not drawn to scale) along with the schematic of a thermopile. (b) Thermography of a large number of people.

The IR thermometers can be broadly classified as:

1. **Spot Infrared (IR) Thermometers:** As discussed above, these thermometers measure the temperature over a spot on the target materials.
2. **Infrared (IR) Scanners:** These devices scan a much larger area than a spot by using a rotating mirror arrangement.
3. **Infrared (IR) Thermal Imaging Cameras:** These devices create a 2D thermal image using temperature measurements at many points over very large areas. The resultant image is commonly known as thermograms as shown in Figure 2.1(b).

During the times of COVID-19, the use of spot infrared (IR) thermometers and IR thermal imaging have been observed. The contactless IR thermometers provide a safe way to measure the temperature of a patient or person by maintaining distance and avoiding contamination of the device. The thermograms are widely used at airports for mass screening of passengers. These images require a powerful backup of hardware and software technologies. The large amount of data continuously generated requires smart technologies such as artificial intelligence (AI) and image processing techniques to accurately monitor and scrutinize the information. This also ensures precise and early detection of persons who may have fever and require further diagnosis.

2.3   SPECTROSCOPY TECHNIQUES

The human body fluids (such as saliva, mucus, urine, and blood plasma) are the primary sources used for the diagnosis of any disease. The use of non-invasive methods for diagnosis of any disorder is becoming more favored due to the simple way of sample collection, and being less painful and less costly. In fact, the diagnosis of COVID-19 is done by collecting nasopharyngeal specimen [1]. The specimen is then analyzed using the polymerase chain reaction (PCR) test. The PCR requires extraction and amplification of the nucleic acid before the analysis. Although the PCR testing is strong and well-established, alternative diagnosing and monitoring methods are being sought, which will provide fast and accurate test results. Of special interest are technologies which would not require the separation of the virus components. In this way, various spectroscopy techniques could prove helpful. This section gives a brief overview about the different light-based technologies, which are currently being used or expected to contribute in viral testing and diagnosing including COVID-19.

2.3.1   PULSE OXIMETER

The COVID-19 patients are routinely checked for vital signs such as body temperature, blood pressure, heartbeat, and oxygen levels. The IR thermometers have proven to be efficient for contactless measuring of the body temperature. The heart rate and the oxygen levels are measured using a pulse oximeter [2]. The pulse oximeter is a small and convenient optical device, which is clipped on the finger of a person, and the results are displayed within seconds. The working of a pulse oximeter can be understood from Figure 2.2. It consists of two light sources of wavelength 650 nm (red light) and 950 nm (IR light). The emitted light from the sources gets partially absorbed by the finger and partially passed through it to reach the detector. The amount of light absorbed by the finger depends on many parameters, and these are used for determining the oxygen saturation values. The hemoglobin carrying oxygen (oxygenated Hb) absorbs more IR light as compared to the red light. On the other hand, the deoxygenated Hb absorbs more red light than IR light. By comparing the ratio between the red and IR light reaching the detector, the oxygen satura- tion in the blood can be determined. Although this is the basic working principle, the actual device is calibrated based on many other factors such as the scattering of light while passing through the finger, and the absorp- tion of light by parts other than the arterial blood. In the case of corona positive patients, a pulse oximeter is a handy device which can timely give information about the oxygen levels in a patient.

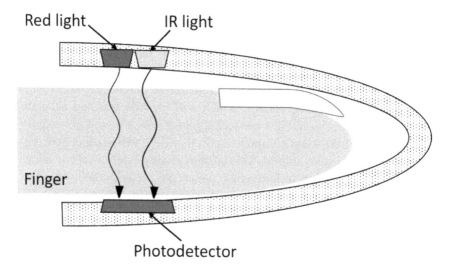

FIGURE 2.2   Simplified diagram of the working of a pulse oximeter.

2.3.2   INFRARED (IR) SPECTROSCOPY

The near-infrared (NIR) spectroscopy makes use of NIR radiations. In terms of biomedical applications, the NIR spectroscopy is gaining interest to be implemented in the testing of viral infections including coronavirus. The NIR spectroscopy is a fast and non-invasive technology, which does not need special sample preparation and reagents for the analysis [3]. The NIR spectrometer measures molecular vibrations in the IR range (700 to 2500 nm), and the asymmetric vibrations corresponding to hydrogen bonds are particularly intense in the IR range. A beam of light is separated into different wavelengths using a diffraction grating. The light strikes the samples and gets absorbed by the material. The absorption in the material depends on the characteristic vibrational frequencies of the constituent molecules. The unabsorbed light passes through the sample and reaches the detector. A comparison between the transmitted light from a reference sample and the test sample gives the required information about the material properties.

In the case of biological samples, the region from 650 to 1000 nm is particularly useful. The water absorbs greatly above 1000 nm and the hemoglobin absorbs significantly below 650 nm. Therefore, this cut-off window is used for the analysis of molecules other than the water and the hemoglobin. For NIR spectroscopy to be successfully implemented for viral detection, it has to be calibrated using pre-acquired reference data. The data sets help in distinguishing the additional signatures introduced in the spectrum due to the virus alone. For this, samples from a large population are required that includes healthy and infected people. The data is then analyzed and standardized using mathematical models. These models may include regression models [3]. It has also been stated that NIR spectroscopy may be used to identify pneumonia in patients based on the water content in the lungs [4]. The lungs of the infected persons show signs of increased fluid or pus in the alveoli, which can be detected based on the change in IR absorption. Similarly, the Fourier transformed infrared spectroscopy (FTIR) is a sensitive instrument for the detection of molecular changes in the cells [5]. This technique has been tested on the analysis of poliovirus and fungal phytopathogens [6, 7], and changes in biomarkers such as lipids, proteins, nucleic acids, and sugars are identified.

The attenuated total reflection Fourier-transform infrared (ATR-FTIR) spectroscopy is another form of vibrational spectroscopy that is useful for the investigation of biological samples and provides results within seconds [8]. The data obtained from these spectrometers are multivariate data, which means there is more than one statistical outcome variable present at a time. To interpret such data, multivariate analysis methods are implemented that make use of mathematical and statistical models.

Some examples of the models are: principal component analysis (PCA), successive projections algorithm (SPA), genetic algorithm (GA), linear discriminant analysis (LDA), and quadratic discriminant analysis (QDA) [9]. The ATR-FTIR spectrum in region from 1800–900 cm^{-1} carries the most significant information about the biological samples. In this spectroscopy technique, the sample is placed on a diamond crystal. The incident IR light passes through the crystal and gets totally internally reflected from the crystal/sample interface (Figure 2.3). This gives rise to evanescent waves that penetrate the sample up to the order of few microns. The absorption (or transmission) of these waves gives the ATR-FTIR spectrum. The spectrum then has to be analyzed in order to determine the characteristic vibrations coming from a particular virus. The spectral peaks are unique depending on the molecular bond that exist in the sample (containing virus). The list of the markers for hepatitis can be found in the literature [10].

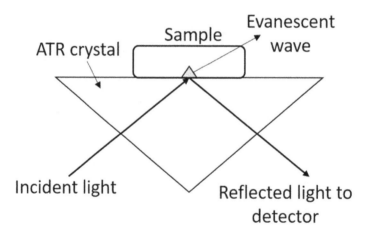

FIGURE 2.3   Schematic representation of the total internal reflection and formation of evanescent wave at the interface between the crystal and the sample.

2.3.3   RAMAN SPECTROSCOPY

Raman spectroscopy is another highly sensitive and useful vibrational spectroscopy technique that allows non-destructive and real-time analysis of biological samples. A Raman spectrum is obtained by the process of scattering of light, whereas in FTIR, it is obtained by absorption of light by the matter. When a monochromatic (laser source) light is incident on the sample, the light may interact with the material either elastically or inelastically. In the elastic scattering, the incident photon is absorbed and reemitted with the same energy (frequency). This is known as Rayleigh scattering. On the other hand, in an inelastic scattering, the absorbed photon may be emitted with frequency higher or lower than the incident photon. The probability of inelastic scattering is very small compared to the Rayleigh scattering. The process of light scattering is shown in Figure 2.4. When the frequency of emitted photon (v_2) is less than the incident frequency (v_1), it is known as Stokes Raman scattering. When the frequency of emitted photon (v_2) is more than the incident frequency (v_1), it is known as anti-Stokes Raman scattering. This phenomenon is known as the Raman effect, and the observed effect is specific to the molecules causing the scattering. Thus, the Raman signals are used for determining the presence of molecules and their states using the inelastic scattering.

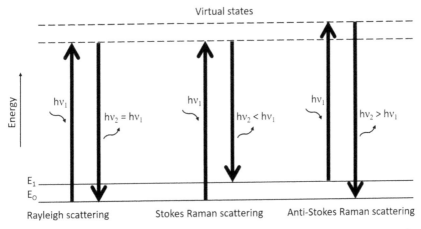

FIGURE 2.4   Illustration of Raman effect demonstrating the elastic (Rayleigh) scattering and inelastic (stokes and anti-stokes) scattering.

Low-frequency Raman spectroscopy has been identified as a useful technique for providing insights into coronavirus detection [11]. The Raman spectroscopy can be used for the analysis of specific viruses or an entire class [11–13]. Valuable insights about the nucleic acids and the structure of the viral proteins can be obtained by Raman analysis [14, 15]. The rapid screening of patients for viral infections have been successfully demonstrated [16], and it is expected that it can also be implemented in case of coronavirus.

2.3.4   *GAS IN SCATTERING MEDIA ABSORPTION SPECTROSCOPY (GASMAS)*

Gas in scattering media absorption spectroscopy (GASMAS) is another technology which is being speculated for use in coronavirus monitoring [4]. The GASMAS may be used to find out the oxygen gas concentration or water in the lungs of a patient. A continuous-wave (CW) laser of desired wavelength is used as a light source. The preferred lasers are semiconductor lasers due to their compact size, low cost, and ease of tuning and light modulation [18]. The GASMAS works on the principle of scattering of light at the interface of different media. When the light enters a sample, it gets scattered due to the inhomogeneity resulting from the gas-filled cavities/pores in the material. The light scatters due to the abrupt change in refractive index of the solid-state material (for example, tissues in case of biological specimen) and the gas. The scattered light is collected by the detector. Now, the absorption traces of free gases are very narrow as compared to the absorption by solid materials. This signature is used to distinguish the signals from gas and the solid material. This way GASMAS can be used to monitor the presence of certain gases including water vapors in the human body. This has been successfully demonstrated in the diagnosis of sinus [18].

2.3.5   *OPTICAL FIBER SENSOR*

A very novel method for rapid detection of COVID-19 has been proposed by Nag et al. [19]. This method is based on the formation of evanescent waves formed during total internal reflection (TIR) of light waves. Optical fiber is a specially designed cable which consists of two transparent materials in the form of co-axial cylinders as shown in Figure 2.5. The inner material (core) has a refractive index greater than the outer material (cladding), which is a necessary requirement for TIR to occur. When a light is incident at an angle greater than the critical angle, it undergoes TIR at the core/cladding interface. At each reflection, some energy is dissipated in the form of evanescent waves. These waves are very sensitive to any change in the refractive index. This dependence can be used as a sensing parameter, and recorded as a change in absorbance. For this, a small portion of the core is exposed by removing the cladding layer and replace it with the testing material. In case of coronavirus detection, the authors proposed two schemes for using this technology. In the first one, the antibodies can be immobilized on gold nanoparticles or on polyaniline-coated optical fibers. If the virus binds to the antibodies, the change in the local refractive index could be noted as a change in intensity of the light coming out of the optical fiber. The second method is based on the immobilization of the virus surface protein for detection of the anti-bodies. Thus, this simple yet powerful technique can be developed for fast screening of infected persons.

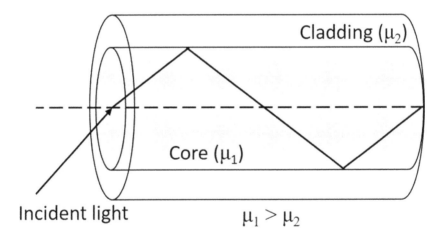

FIGURE 2.5   The total internal reflection in an optical fiber cable.

2.4   IMAGING TECHNIQUES

The current imaging techniques like X-ray images and computed tomography (CT) are quite useful and well-established in diagnosing many illnesses. In order to understand the role of these technologies, a short description of the physical principles involved is given. Thereafter, the case studies related to COVID-19 are presented.

X-rays are radiations having wavelength in the range 0.01 to 10 nm, which means they have energies much higher than the visible range. As a result, the X-rays can penetrate through solid, liquid, and gases. The amount of penetration of X-rays depends on the material properties. When X-rays are passed through the human body, the low energy radiations are absorbed by the high-density parts of the body. The rest of the X-rays pass through and strike a photographic plate. The X-rays form a negative image on the photographic plate. To elucidate this, consider the bones and the flesh of the human body. The bones absorb the X-rays more than the flesh due to the presence of calcium in them. However, the X-ray images shows the bones as shades of white, whereas the rest of the image is dark.

CT also uses X-ray radiations like X-ray imaging, although the working principle is different in the two cases. In CT scans, cross-sectional images of the body are formed based on the X-ray interaction with the body. As stated earlier, the X-rays penetrate the body to different extent depending on the composition and density of the interacting material. While passing through a medium, all radiations tend to attenuate with distance. The attenuation coefficient defines the amount of radiations absorbed per unit thickness in a medium. The CT scans are constructed based on the measurement of the attenuation coefficients of the X-rays in the body. During a CT scan, the source and the detector are rotated around the body that results in capturing a 3D image of the interior rather than 2D image obtained in X-ray imaging. The image formation is done using complicated mathematical models.

There are several reports which look into the use of imaging techniques to reveal the information about the coronavirus. COVID-19 infected cases portray irregular distribution and shading in the chest images [20]. Shah et al. [21] discussed that the chest X-rays and CT scans of COVID-19 patients do reveal some opacities and signs of infection, but its distinction with other forms of infection (like pneumonia) cannot be ascertained. Moreover, the same features are observed at the early stages and advanced stages of the infection [22], making it hard to correlate with the condition of the patient. Thus, alone X-rays and CT scans may not be sufficient for declaring the results of coronavirus infection and tests like PCR are required along with it. In order to increase the likeliness of correct predictions from the chest X-ray images, Ozturk et al. [22] have proposed a machine-learning algorithm. The authors analyzed the images of six different cases using hand-crafted feature extraction method: No finding,

COVID, pneumocystis-pneumonia, ARds, Sars, and streptococcus classes. A 90% accuracy was obtained in predicting the results from the images for diagnosis purposes. However, the authors notified that since the COVID-19 dataset was limited at the time of publication, further studies may be needed in this direction.

2.5 BIOSENSORS

The use of biosensors for the rapid detection of coronavirus has also been discussed by several researchers, which will be discussed in this section. A biosensor is a device which uses a transducer and a biological matter such as enzymes, nucleic acids, or antibodies to detect the presence of the substance under investigation. The testing material (analyte) reacts with the biological matter, causing a certain change (for example, resistance) that is converted into an electrical signal by the transducer (Figure 2.6). Various chip-based and paper-based biosensors are envisioned to be helpful for rapid diagnosis of viral infections [23]. The advantages of these biosensors are their low cost, fast testing speed, and simple operational procedure. These biosensors are based on the detection of nucleic acids, antigens, or antibodies from various samples such as blood, saliva, sputum, and nasal secretion [24–26]. The chip-based biosensors are commonly made using polydimethylsiloxane (PDMS) or poly(methyl methacrylate) (PMMA) [27, 28]. The chip-based biosensors involve multiple channels for automated extraction of nucleic acid and isothermal amplification. Further components for real-time signal detection are based on fluorescence, colorimetry, or electrochemical detection [23]. The paper-based biosensors are gaining more interest than chip-based because of several advantages they offer. The paper-based sensors are biodegradable, have lower cost, are easy to fabricate and chemically modify [29–31]. The lateral flow test strips have been put to test for COVID-19 detection. They detect the antibodies in the patient sample [24, 26].

Plasmonic biosensors may offer reliable detection of COVID-19 virus as well. The localized surface plasmon resonance (LSPR)-based sensors have been used to detect clinical analytes [32]. The LSPR refers to the coherent oscillations induced by the photons in the conducting surface electrons. This localized plasmonic resonance can be modulated as a result of change in certain properties such as refractive index and molecular binding. A dual functional plasmonic biosensor was tested by Qiu et al. [33]. The device was capable of inducing plasmonic photothermal (PPT) effect by using gold nanoparticle islands. The nanoparticles strongly absorb the photons and cause localized heating due to non-radiative emission. This phenomenon is used for the hybridization of the nucleic acid strands. The LSPR is then used for the detection of the COVID-19 virus.

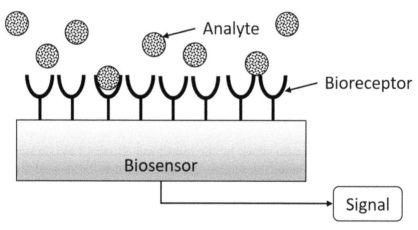

FIGURE 2.6   Schematic illustration of a biosensor.

The importance of graphene in the fight against COVID-19 has also been reflected upon by considering its uses in several forms. Due to the monolayered carbon structure, graphene possesses high surface area, which is well exposed. The high surface area is advantageous for detecting even a single molecule due to a change in its electrical properties. Thus, graphene serves as an ideal choice for the construction of sensors [34]. The graphene surface can be easily functionalized according to the requirements. The oxygen present on the surface of graphene oxide also provides reactive sites for nucleic acids, enzymes, or proteins [35]. The virus proteins can be detected by integrating antibodies on the surface of graphene oxide. It has been demonstrated that sulfonated magnetic nanoparticles functionalized with graphene oxide can capture herpes simplex virus type 1, after which it can be destroyed by using NIR light [36]. Similar attempts can be made to capture and destroy the COVID-19 virus in the future.

A field-effect transistor (FET) biosensor using graphene have been developed which shows promising results in the detection of COVID-19 causing virus [37]. The fabricated FET uses graphene coated with anti-bodies using PBASE against spike protein of the virus. The performance of the biosensor was analyzed under multiple conditions with successful outcomes. The loading of antibodies was essential as the biosensor did not give any signals without the immobilized antibodies. The biosensor was able to detect the antigen protein of the virus. The protein could also be detected when the analyte contained the universal transport medium (UTM) in which the nasopharyngeal swabs are suspended. Further, the FET response was obtained for cultured virus and nasopharyngeal swabs containing the virus in the UTM. The graphene-based FET biosensor was able to successfully distinguish between non-infected and infected swabs, and also between COVID-19 causing virus and MERS-CoV spike protein.

2.6   SANITIZATION AND IDENTIFICATION THROUGH ENVIRONMENTAL MONITORING

2.6.1   UV-TECHNOLOGY FOR SANITIZATION

The COVID-19 infection is rapidly transmitted if a person comes in contact with surroundings containing the virus. The mode of transmission may be direct inhalation of the droplets exhaled by an infected person in the form of cough, sneeze, or breath. It may also be transmitted through contact with any contaminated surface. In order to suppress the spread of the virus, many guidelines have been issued from time to time. One of the important strategies include frequent sanitization of all the surrounding environment. It has been investigated that the coronavirus was found on the surfaces of counters, bed frames, doorknobs, floors, bathrooms, shoe soles, stationary items, money, car interiors, etc. [38]. This requires an extended effort to sanitize all objects for prevention of the coronavirus spread. The standard protocol of using disinfectants might not be feasible in all cases, especially where money and important documents are concerned. Moreover, the sanitization of large spaces and a huge number of items is a laborious task. Therefore, alternate methods need to be devised which can sanitize large areas in a small time and in an effective manner.

The inactivation of microbes has been long done using ultraviolet (UV) light during water purification, medical sterilization, and food processing [39]. A similar technique can be used to disinfect spaces and equipment from the coronavirus. The sterilization technique being developed is based on using UV germicidal irradiation (UVGI), which uses UV-C rays with wavelength ranging from 100–280 nm. However, the UV-C radiations might affect the polymer material used for making personal protective equipment (PPE) such as N95 respirators, and thus compromise its efficiency. As an alternative, UV-B with a wavelength range of 280–315 nm may be used. However, the work on using UV sterilization for coronavirus and PPE is still on examination stage. Nonetheless, the coronavirus has been shown to be sensitive to the exposure of UV light [40]. After the technique has been successfully tested, it may be possible to use the UV machine to sanitize large areas in a very short time. A team based in the USA is making a handheld UV device, which could be used in hospitals, homes, shops, restaurants, etc., to destroy the coronavirus [41]. It was demonstrated that 99.9% of the virus could be destroyed within 30 seconds. Thus, the use of UV light for disinfecting large areas would assist in controlling the transmission of the virus.

In order to develop the mechanism of disinfection using UV light, the technology involved in the process needs to be understood. The UV-C used in the UVGI is produced by a lamp which emits short-wavelength radiations. The conventional lamps are usually made of mercury vapors enclosed in a fused quartz tube. The mercury vapors are either at low or high pressure depending on the requirement. The low-pressure lamps are similar to fluorescent lamps, but do not contain fluorescent phosphorus. Moreover, the quartz tube is used instead of the glass because the latter absorbs the UV radiations. Whereas the low-pressure mercury lamps give off 253.7 nm radiations, a broader emission may be obtained by using high-pressure mercury lamps, which work on the principle of arc discharge lamps. Since mercury is toxic in nature, alternative technology such as excimer lamps and light-emitting diodes (LEDs) are being used. In an excimer lamp, diatomic molecules are used which are excited from the ground state to the excited state using electric discharge method. The excited electrons come back spontaneously to the ground state emitting photons in the UV region. A wide range of radiation lamps may be made using different excimers. The LED technology is also mercury-free and uses solid-state semiconductor materials such as aluminum gallium nitride (AlGaN). The semiconductor LEDs are tunable for specific needs by doping or chemical modification of the active material. The advantages of LED lamps lie in its compact size; however, much work is needed to optimize its efficiency, device lifetime, and cost as compared to conventional methods.

The UV light gets absorbed in the RNA and DNA of the cells, and thereby damages the DNA of microorganisms (virus, bacteria, and pathogens) by which they lose their ability to replicate [40]. This inactivates the microorganisms and further reproduction of them, thus, controlling the spread of it. The intensity, wavelength of UV light, and exposure time are the key parameters which needs to be adjusted for the disinfection of different microorganisms. UV light is also effective against some drug-resistant bacteria as well. It provides a non-chemical method to disinfect the objects including water and air. Moreover, there has not been any indicators that the microbes develop resistance to light damage, as is the case with chemicals. Thus, UV disinfection method could prove to be vital in controlling the spread of coronavirus once the approved dosage and other technical factors have been tested successfully.

2.6.2   *LARGE AREA ENVIRONMENTAL MONITORING*

The hardship of COVID-19 has made clear the urgency of large-scale monitoring of infected persons, so that the spread can be controlled in a timely manner. The technique that has been used so far relies only on increasing the number of tests and declaring containment zones in those areas where there is an increase in the number of positive cases. However, it is hard to determine before testing the possibility of a locality to be infected. There should be certain ways in which primary stage screening can be done to detect the hotspots at the very outset of the spread. Previous studies have shown that there is a high probability of coronavirus to survive in wastewater [42]. Therefore, the monitoring of wastewater is a suitable technique for identifying possible hotspots in a locality. Further, it is also required that the wastewater may be effectively treated to make it infection-free. For this, the use of nanofiber filters have been proposed [42].

It has been reported that the fecal matter of infected human beings contains coronavirus [43, 44]. This poses the threat of infection spreading through wastewaters, and hence requires measures to ensure proper treatment of wastewater. As discussed in the previous sub-section, UV radiations are effective in destroying coronavirus and thus can be utilized to treat large amounts of water in a short span. Moreover, Venugopal et al. [42] proposed electro-spun nanofiber membrane to capture coronavirus (CoVs) from wastewater, which could then be analyzed for possible contamination. This way areas with CoVs could be targeted for faster isolation and containment. Such an action was taken in The Netherlands by taking wastewater samples and testing them before and after treatment [45]. It was found that the untreated water tested positive, while the treated water contained viral RNA. However, it could not be established whether the virus was infectious or not. The wastewater-based epidemiology (WBE) model has been previously implemented to monitor and control polio and hepatitis A [46]. Hence, such technique could prove to be useful for a macro-level analysis of the spread of an infection.

2.7   3D PRINTING TECHNOLOGY FOR OVERCOMING SHORTAGE OF SUPPLY

Three-dimensional printing technology (3DPT) is inspired from the conventional printer technology. Whilst the conventional ink printers produce a two-dimensional pattern on any printable surface, the 3D printers extend this printing to three dimensions, creating a 3D object. 3D printing has become a method to manufacture goods using different materials as the "ink." A vast range of materials are used as an input material in 3DPT, and further research to include other materials for 3D printing is being carried out. The 3D printing materials include polymers, resins, metals, alloys, ceramics, and concrete [47]. The 3DPT finds wide application in various domains such as automobile, aerospace, biomedical, construction, and textile industry [48–51]. The advantages of 3DPT are the flexibility to design the objects of complex geometries, customization of the shapes and sizes, reduction of costs and time, and fast production of the goods.

The first step in 3D printing technique is developing a 3D model of the object to be produced. This model is made using computer-aided design and software. Secondly, the design is converted into a layered format understood by the 3D printer. This process is referred to as slicing, and it is a method to divide the 3D model into hundreds of horizontal layers. Thereafter, the sliced model file is transferred to the 3D printer, which prints (deposits) the material layer-by-layer to produce the complete object [52]. Since the product is developed based on the computer code entered into the printer, the difficulty of molds/templates required for a particular design is overcome, and this further reduces the cost while ensuring greater customization.

The use of 3D materials has increased in the medical field over the past years. Initially, 3D printing was used to manufacture medical items like hearing aids and prosthetics [53]. Thereafter, 3D printing has been explored for producing and transplanting engineered tissues and organs [54, 55]. Other examples include drug delivery systems [56], fabrication of bones [57], cartilages [58], and dental implants [59]. The COVID-19 pandemic has created a new outlook when it comes to using 3D printing technology. The shortage of medical devices that was encountered all over the world led to exploring the possibility of fast production of medical tools. The face masks, face shields, PPE, and testing kits are some of the medical goods that are extensively used in medical emergencies. The production of these items, along with some others (like ventilator valves, nasal swabs, oxygen masks, door opener, etc.), can be realized through the use of 3DPT [60, 61]. The fitting of the masks on the face is an essential requirement for the effectiveness of the mask in preventing spread of the virus. Therefore, a customization of the mask can be achieved through the accurate modeling of the customer's facial features. Although the use of 3DPT has been seen as an opportunity to develop products during COVID-19, the safety standards and other procedures also need to be taken care of that will ensure whether the 3D printed products pass the quality check of the approved guidelines.

Many companies adopted 3DPT to supply equipment to the hospitals during COVID-19. A brief understanding of the different techniques and materials used for 3DPT is essential to appreciate its use in the present circumstances and future occurrences. The 3DPT can be classified into four techniques:

1. **Fused Deposition Modeling (FDM):** In this technique, the material is melted and directly deposited layer-by-layer based on the 3D diagram provided through CAD. This is the most famous and simple technique. The preferable materials are polymers [62].
2. **Selective Laser Sintering/Melting (SLS/SLM):** In this process, thin layers of closely packed fine powder is fused together by irradiating it with laser. This process is continued until the required 3D object is made. SLS can be used for materials ranging from

metals, alloys, and polymers, while SLM is preferably used for metals and alloys [47].

3. **Stereolithography (SLA):** In this method, an electron beam or UV light is used to induce a chain reaction in the layer of resin or monomer material. This results into the production of a polymer, which after being partly solidified is used for assembling subsequent layers [63].

4. **Digital Light Processing (DLP):** This process uses projected light to initiate polymerization of the material to form pre-designed structures [63]. The materials of choice include photosensitive resins, ceramic filled materials, and metals.

The components of a face mask are made of different materials. For example, the nose clip may be made of a metal (aluminum), and it can be 3D printed using SLS/SLM technique. The layers of N95 mask can also be deposited one by one. Similarly, the ear straps can be made using thermoplastic elastomers through SLS or FDM methods. Other parts such as nose foam (polyurethane), face seal (acrylonitrile butadiene styrene (ABS)), and exhalation valve (polyvinyl chloride (PVC)) can be printed by using different materials and different techniques [61].

The transparent face shields are composed of polymers such as PVC and other synthetic polymers [17]. These materials are highly transparent and lightweight, and can be deposited using 3DPT. The nasal swabs used during sample collection for corona test can be efficiently made using 3DPT. This would require flexible polymer such as polystyrene for the stick, and the bud can be made using calcium hydrogels [17]. Other than these, the ventilator valves, medicines, microscopes, and safety goggles also saw potential use of 3DPT during the pandemic [55, 61]. Thus, the 3DPT can mass-produce the necessary items locally without much dependency on the global supply chain [55]. The on-demand supply of equipment ensures greater control over the spread of the disease in a timely manner.

2.8   CONCLUSION

The COVID-19 pandemic is a real concern, and it needs to be acutely studied so that humans are better prepared for future recurrences of the same virus or other viruses. The detailed investigation of the virus and its effects require the contribution from every field of research. The materials science, biomedical, engineering, mathematics, environmental science, and even social sciences have contributed to understanding the grave effects of this pandemic. Although substantial improvement has been made in the detection and management of the virus spread, there is still a lot of work that needs to be done. The handling of any epidemic or pandemic situation can be divided into several parts such as: the detection, the screening, the treatment, the safety, the environmental protection, and the socio-economic effects. It becomes very important for all the domains to contribute towards the research of viral infections. The past knowledge from similar viral spreads contributes greatly towards understanding the current situation. Most of the technologies discussed in this chapter were based on the studies that had been previously done on viruses, bacteria, and other biological organisms. Thus, it is evident that optoelectronic devices, biosensors, and vibrational spectroscopy have a huge role in the pandemic. It is only required to further investigate these technologies and customize it based on the properties of the COVID-19 causing virus. Further, the development of fast-computing methods and models have also made it possible for mass screening and detection. Techniques like image processing by using AI can also contribute in smart detection of the infection with higher accuracy and speed.

KEYWORDS

- 3D printing technology
- biosensors
- COVID-19
- linear discriminant analysis
- successive projections algorithm
- thermal screening
- vibrational spectroscopy

REFERENCES

1. Piras, A., et al., (2020). Nasopharyngeal swab collection in the suspicion of COVID-19. *American Journal of Otolaryngology-Head and Neck Medicine and Surgery,* *41*(5), 102551. doi: 10.1016/j.amjoto.2020.102551.

2. Chan, E. D., Chan, M. M., & Chan, M. M., (2013). Pulse oximetry: Understanding its basic principles facilitates appreciation of its limitations. *Respiratory Medicine, 107*(6), 789–799. W. B. Saunders. doi: 10.1016/j.rmed.2013.02.004.

3. Sakudo, A., Suganuma, Y., Kobayashi, T., Onodera, T., & Ikuta, K., (2006). Near-infrared spectroscopy: Promising diagnostic tool for viral infections. *Biochemical and Biophysical Research Communications, 341*(2), 279–284. Elsevier. doi: 10.1016/j.bbrc.2005.12.153.

4. Nogueira, M. S., (2020). Biophotonic telemedicine for disease diagnosis and monitoring during pandemics: Overcoming COVID-19 and shaping the future of healthcare. *Photodiagnosis and Photodynamic Therapy, 31*, 101836. Elsevier B.V. doi: 10.1016/j.pdpdt.2020.101836.

5. Erukhimovitch, V., Talyshinsky, M., Souprun, Y., & Huleihel, M., (2005). FTIR microscopy detection of cells infected with viruses. *Methods in Molecular Biology (Clifton, N.J.), 292*, 161–172. doi: 10.1385/1-59259-848-x:161.

6. Salman, A., Tsror, L., Pomerantz, A., Moreh, R., Mordechai, S., & Huleihel, M., (2010). FTIR spectroscopy for detection and identification of fungal phytopathogens. *Spectroscopy, 24*(3, 4), 261–267. doi: 10.3233/SPE-2010-0448.

7. Lee-Montiel, F. T., Reynolds, K. A., & Riley, M. R., (2011). Detection and quantification of poliovirus infection using FTIR spectroscopy and cell culture. *Journal of Biological Engineering, 5*, 16. doi: 10.1186/1754-1611-5-16.

8. Das, C. E., & Nogueira, M. S., (2020). Optical techniques for fast screening - Towards prevention of the coronavirus COVID-19 outbreak. *Photodiagnosis and Photodynamic Therapy, 30*, 101765. Elsevier B.V. doi: 10.1016/j.pdpdt.2020.101765.

9. Santos, M. C. D., Morais, C. L. M., & Lima, K. M. G., (2020). ATR-FTIR spectroscopy for virus identification: A powerful alternative. *Biomedical Spectroscopy and Imaging*, 1–16. doi: 10.3233/bsi-200203.

10. Roy, S., Perez-Guaita, D., Bowden, S., Heraud, P., & Wood, B. R., (2019). Spectroscopy goes viral: Diagnosis of hepatitis B and C virus infection from human sera using ATR-FTIR spectroscopy. *Clinical Spectroscopy, 1*, 100001. doi: 10.1016/j.clispe.2020.100001.

11. Jacobi, L., (2020). Low-frequency Raman spectroscopy as a diagnostic tool for COVID-19 and other coronaviruses. *Royal Society Open Science.*

12. Blanch, E. W., et al., (2002). Molecular structures of viruses from Raman optical activity. *Journal of General Virology, 83*(10), 2593–2600. doi: 10.1099/0022-1317-83-10-2593.

13. Cialla, D., et al., (2010). TERS as a diagnostic tool for single virus detection. In: *AIP Conference Proceedings* (Vol. 1267, No. 1, pp. 1269–1270). doi: 10.1063/1.3482418.

14. Němeček, D., & Thomas, G. J., (2009). Raman spectroscopy of viruses and viral proteins. In: *Frontiers of Molecular Spectroscopy* (pp. 553–595). Elsevier.

15. Němeček, D., & Thomas, G. J., (2009). Raman spectroscopy in virus structure analysis. In: *Digital Encyclopedia of Applied Physics*. Weinheim, Germany: Wiley-VCH Verlag GmbH & Co. KGaA.

16. Mahmood, T., et al., (2018). Raman spectral analysis for rapid screening of dengue infection. *Spectrochimica Acta - Part A: Molecular and Biomolecular Spectroscopy, 200*, 136–142. doi: 10.1016/j.saa.2018.04.018.

17. Bachtiar, E. O., et al., (2020). 3D printing and characterization of a soft and biostable elastomer with high flexibility and strength for biomedical applications. *Journal of the Mechanical Behavior of Biomedical Materials, 104*, 103649. doi: 10.1016/j. jmbbm.2020.103649.

18. Svanberg, S., (2013). Gas in scattering media absorption spectroscopy - from basic studies to biomedical applications. *Laser & Photonics Reviews, 7*(5), 779–796. doi: 10.1002/lpor.201200073.

19. Nag, P., Sadani, K., & Mukherji, S., (2020). Optical fiber sensors for rapid screening of COVID-19. *Transactions of the Indian National Academy of Engineering, 5*(2), 233–236. doi: 10.1007/s41403-020-00128-4.

20. Wang, D., et al., (2020). Clinical characteristics of 138 hospitalized patients with 2019 novel coronavirus-infected pneumonia in Wuhan, China. *JAMA - Journal of the American Medical Association, 323*(11), 1061–1069. doi: 10.1001/jama.2020.1585.

21. Shah, K., Kamler, J., Phan, A., & Toy, D., (2020). Imaging & other potential predictors of deterioration in COVID-19. *American Journal of Emergency Medicine, 38*(7), 1547.e1–1547.e4, doi: 10.1016/j.ajem.2020.04.075.

22. Ozturk, S., Ozkaya, U., & Barstugan, M., (2020). *Classification of Coronavirus Images Using Shrunken Features*. doi: 10.1101/2020.04.03.20048868.

23. Choi, J. R., (2020). Development of point-of-care biosensors for COVID-19. *Frontiers in Chemistry, 8*. doi: 10.3389/fchem.2020.00517.

24. Li, Z., et al., (2020). Development and clinical application of a rapid IgM-IgG combined antibody test for SARS-CoV-2 infection diagnosis. *Journal of Medical Virology, 92*(9), 1518–1524. doi: 10.1002/jmv.25727.

25. Zhifeng, J., Feng, A., & Li, T., (2020). Consistency analysis of COVID-19 nucleic acid tests and the changes of lung CT. *Journal of Clinical Virology, 127*. doi: 10.1016/j.jcv.2020.104359.

26. Sheridan, C., (2020). Fast, portable tests come online to curb coronavirus pandemic. *Nature Biotechnology, 38*(5), 515–518. NLM (Medline). doi: 10.1038/ d41587-020-00010-2.

27. Zhang, L., Ding, B., Chen, Q., Feng, Q., Lin, L., & Sun, J., (2017). Point-of-care-testing of nucleic acids by microfluidics. *TrAC - Trends in Analytical Chemistry, 94*, 106–116. Elsevier B.V. doi: 10.1016/j.trac.2017.07.013.

28. Zarei, M., (2017). Advances in point-of-care technologies for molecular diagnostics. *Biosensors and Bioelectronics, 98*, 494–506. Elsevier Ltd. doi: 10.1016/j. bios.2017.07.024.

29. Hu, J., et al., (2017). Multiple test zones for improved detection performance in lateral flow assays. *Sensors and Actuators, B: Chemical, 243*, 484–488. doi: 10.1016/j. snb.2016.12.008.

30. Böhm, A., Trosien, S., Avrutina, O., Kolmar, H., & Biesalski, M., (2018). Covalent attachment of enzymes to paper fibers for paper-based analytical devices. *Frontiers in Chemistry, 6*. doi: 10.3389/fchem.2018.00214.

31. Choi, J. R., Yong, K. W., Choi, J. Y., & Cowie, A. C., (2019). Emerging point-of-care technologies for food safety analysis. *Sensors (Switzerland), 19*(4). MDPI AG. doi: 10.3390/s19040817.

32. Haes, A. J., Chang, L., Klein, W. L., & Van, D. R. P., (2005). Detection of a biomarker for Alzheimer's disease from synthetic and clinical samples using a nanoscale optical

biosensor. *Journal of the American Chemical Society, 127*(7), 2264–2271. doi: 10.1021/ja044087q.

33. Qiu, G., Gai, Z., Tao, Y., Schmitt, J., Kullak-Ublick, G. A., & Wang, J., (2020). Dual-functional plasmonic photothermal biosensors for highly accurate severe acute respiratory syndrome coronavirus 2 detection. *ACS Nano, 14*(5), 5268–5277. doi: 10.1021/acsnano.0c02439.

34. Chauhan, N., Maekawa, T., & Kumar, D. N. S., (2017). Graphene based biosensors - Accelerating medical diagnostics to new dimensions. *Journal of Materials Research, 32*(15), 2860–2882. Cambridge University Press. doi: 10.1557/jmr.2017.91.

35. Palmieri, V., & Papi, M., (2020). Can graphene take part in the fight against COVID-19? *Nano Today, 33*, 100883. doi: 10.1016/j.nantod.2020.100883.

36. Deokar, A. R., et al., (2017). Graphene-based 'hot plate' for the capture and destruction of the herpes simplex virus type 1. *Bioconjugate Chemistry, 28*(4), 1115–1122. doi: 10.1021/acs.bioconjchem.7b00030.

37. Seo, G., et al., (2020). Rapid detection of COVID-19 causative virus (SARS-CoV-2) in human nasopharyngeal swab specimens using field-effect transistor-based biosensor. *ACS Nano, 14*(4), 5135–5142. doi: 10.1021/acsnano.0c02823.

38. *Cleaning and Disinfection for Households | CDC.* https://www.cdc.gov/coronavirus/2019-ncov/prevent-getting-sick/cleaning-disinfection.html (accessed on 22 June 2021).

39. Heßling, M., Hönes, K., Vatter, P., & Lingenfelder, C., (2020). Ultraviolet irradiation doses for coronavirus inactivation - review and analysis of coronavirus photoinactivation studies. *GMS Hygiene and Infection Control, 15*, p. Doc08. doi: 10.3205/dgkh000343.

40. Lindblad, M., Tano, E., Lindahl, C., & Huss, F., (2020). Ultraviolet-C decontamination of a hospital room: Amount of UV light needed. *Burns, 46*(4), 842–849. doi: 10.1016/j.burns.2019.10.004.

41. Bradley, D., (2020). Shedding ultraviolet light on coronavirus. *Materials Today, 37*, 6, 7. doi: 10.1016/j.mattod.2020.05.007.

42. Venugopal, A., et al., (2020). Novel wastewater surveillance strategy for early detection of coronavirus disease 2019 hotspots. *Current Opinion in Environmental Science and Health, 17*, 8–13. doi: 10.1016/j.coesh.2020.05.003.

43. Xu, Y., et al., (2020). Characteristics of pediatric SARS-CoV-2 infection and potential evidence for persistent fecal viral shedding. *Nature Medicine, 26*(4), 502–505. doi: 10.1038/s41591-020-0817-4.

44. Wu, Y., et al., (2020). Prolonged presence of SARS-CoV-2 viral RNA in faecal samples. *The Lancet Gastroenterology and Hepatology, 5*(5), 434, 435. Elsevier Ltd. doi: 10.1016/S2468-1253(20)30083-2.

45. Lodder, W., & De Roda, H. A. M., (2020). SARS-CoV-2 in wastewater: Potential health risk, but also data source. *The Lancet Gastroenterology and Hepatology, 5*(6), 533, 534. Elsevier Ltd. doi: 10.1016/S2468-1253(20)30087-X.

46. Hart, O. E., & Halden, R. U., (2020). Computational analysis of SARS-CoV-2/COVID-19 surveillance by wastewater-based epidemiology locally and globally: Feasibility, economy, opportunities and challenges. *Science of the Total Environment, 730*, 138875. doi: 10.1016/j.scitotenv.2020.138875.

47. Ngo, T. D., Kashani, A., Imbalzano, G., Nguyen, K. T. Q., & Hui, D., (2018). Additive manufacturing (3D printing): A review of materials, methods, applications and challenges. *Composites Part B: Engineering, 143,* 172–196. doi: 10.1016/j.compositesb.2018.02.012.

48. Chakraborty, S., & Biswas, M. C., (2020). 3D printing technology of polymer-fiber composites in textile and fashion industry: A potential roadmap of concept to consumer. *Composite Structures, 248,* 112562. Elsevier Ltd. doi: 10.1016/j.compstruct.2020.112562.

49. Tay, Y. W. D., Panda, B., Paul, S. C., Noor, M. N. A., Tan, M. J., & Leong, K. F., (2017). 3D printing trends in building and construction industry: A review. *Virtual and Physical Prototyping, 12*(3), 261–276. Taylor and Francis Ltd. doi: 10.1080/17452759.2017.1326724.

50. Ahangar, P., Cooke, M. E., Weber, M. H., & Rosenzweig, D. H., (2019). Current biomedical applications of 3D printing and additive manufacturing. *Applied Sciences, 9*(8), 1713. doi: 10.3390/app9081713.

51. Duda, T., & Raghavan, L. V., (2016). 3D metal printing technology. *IFAC-PapersOnLine, 49*(29), 103–110. doi: 10.1016/j.ifacol.2016.11.111.

52. Chua, C. K., Leong, K. F., & Lim, C. S., (2010). *Rapid Prototyping: Principles and Applications, Third Edition.* World Scientific Publishing Co.

53. Berman, B., (2012). 3-D printing: The new industrial revolution. *Business Horizons, 55*(2), 155–162. doi: 10.1016/j.bushor.2011.11.003.

54. Murphy, S. V., & Atala, A., (2014). 3D bioprinting of tissues and organs. *Nature Biotechnology, 32*(8), 773–785. Nature Publishing Group. doi: 10.1038/nbt.2958.

55. Attaran, M., (2020). 3D printing role in filling the critical gap in the medical supply chain during COVID-19 pandemic. *American Journal of Industrial and Business Management, 10*(05), 988–1001. doi: 10.4236/ajibm.2020.105066.

56. Goole, J., & Amighi, K., (2016). 3D printing in pharmaceutics: A new tool for designing customized drug delivery systems. *International Journal of Pharmaceutics, 499*(1, 2), 376–394. Elsevier B.V. doi: 10.1016/j.ijpharm.2015.12.071.

57. Wen, Y., et al., (2017). 3D printed porous ceramic scaffolds for bone tissue engineering: A review. *Biomaterials Science, 5*(9), 1690–1698. Royal Society of chemistry. doi: 10.1039/c7bm00315c.

58. De Mori, A., Peña, F. M., Blunn, G., Tozzi, G., & Roldo, M., (2018). 3D printing and electrospinning of composite hydrogels for cartilage and bone tissue engineering. *Polymers, 10*(3), 285. doi: 10.3390/polym10030285.

59. Tunchel, S., Blay, A., Kolerman, R., Mijiritsky, E., & Shibli, J. A., (2016). 3D printing/additive manufacturing single titanium dental implants: A prospective multicenter study with 3 years of follow-up. *International Journal of Dentistry, 2016.* doi: 10.1155/2016/8590971.

60. Brohi, S. N., Jhanjhi, N., Brohi, N. N., & Brohi, M. N., (2020). *Key Applications of State-of-the-Art Technologies to Mitigate and Eliminate COVID-19, 2019.* doi: 10.36227/TECHRXIV.12115596.V2.

61. Ishack, S., & Lipner, S. R., (2020). Applications of 3D printing technology to address COVID-19-related supply shortages. *American Journal of Medicine, 133*(7), 771–773. doi: 10.1016/j.amjmed.2020.04.002.

62. Mohamed, O. A., Masood, S. H., & Bhowmik, J. L., (2015). Optimization of fused deposition modeling process parameters: A review of current research and future prospects. *Advances in Manufacturing, 3*(1), 42–53. doi: 10.1007/s40436-014-0097-7.
63. Zhang, J., Hu, Q., Wang, S., Tao, J., & Gou, M., (2020). Digital light processing based three-dimensional printing for medical applications. *International Journal of Bioprinting, 6*(1), 12–27. doi: 10.18063/ijb.v6i1.242.

CHAPTER 3

Digital and Personalized Healthcare System for COVID-19 and Future Pandemics

VARSHA SINGH

Centre for Life Sciences, Chitkara School of Health Sciences, Chitkara University, Punjab, India, E-mail: varsha.singh@chitkara.edu.in

ABSTRACT

It has become clear over the decade that our health is not that deterministic. Clinicians must measure various nucleotides to decide on our care and incorporate a growing number of health data generated and tracked by people. The chapter addresses important aspects of personalizing and digital health diagnostics for incorporating data in the health care system, so that informed choices can be made, as the idea of 'one pill for all' does not encompass all. The benefits of using predictive analytics are the same as many digital health categories: better care and lower costs. The difference is that the path to achieving these benefits is only possible through the use of these technologies—through personalized attention. It is essential and real that emphasis is reduced to a set of algorithmically generated probabilities. But it is also the dream. Predictive analytics is the learning process from historical data to make (or any unknown) predictions for the future. For health care, the predictive analysis will allow the right decisions to be made so that each patient can receive personalized care. The chapter will discuss how digital health and personalized analytics directly affect patient care and prepare us for future COVID-19-like pandemics. Big data and the development of algorithms have generated interest and excitement in predictive analysis. Fortunately, a wave of modern technology, including many open-source technologies, has emerged in the past few years to process and manage all these healthcare data. The importance of having a history around an individual's historical data is so crucial, particularly in health care, when we still know what is "healthy" or "standard." Recommendations often provide ranges for each patient that are incorporated in clinical alert systems without any context. Predictive analytics can be used to improve predictive certainty. Personalized treatment can result from high-trust algorithms that can predict workable interventions to improve long-term health outcomes of individuals affected with COVID-19 and prepare the healthcare system for future pandemics. Nevertheless, it is not an easy task to build digital models that can detect symptoms in patients and choose personalized treatment for COVID-19 like pandemics; however, it should involve numerous factors including the ability to comprise new types and sources of information, accurate, robust predictive models, timely data, prediction accuracy, simplicity, and contextual recommendations. Hence, healthcare should be highly personalized and individual-centric for therapeutic purposes. The prescribed medication to ensure adequate control depends on the severity of the condition and the frequency of the causes at individual levels and should not be prescribed blindly. COVID-19 is one example that requires disease management and also requires complete personalization of treatment in the context of causes and vulnerabilities. This being the main reason, leads to active involvement of various stakeholders such as researchers, clinicians, digital, and IoT experts, and pharmaceutical experts to collaborate and build a digitized and personalized model for detecting and treating the pandemic.

> "…can empower consumers to make better-informed decisions about their own health and provide new options for facilitating prevention, early diagnosis of life-threatening diseases, and management of chronic conditions outside of traditional care settings."
>
> *—Digital Health Definition,*
> *United States Food and Drug Administration*

3.1 INTRODUCTION

In the past decade, pandemic outbreaks, such as severe acute respiratory syndrome-corona virus (SARS-CoV), middle east respiratory syndrome (MERS)-related coronavirus (CoVs), and the recent SARS-CoV-2 causing Coronavirus diseases (COVID-19), marked a pivotal moment for healthcare system. Pandemic control requires for an unequaled need to use existing medical knowledge, technology, and patient databases to integrate human biology and revolutionize next-generation healthcare. Individualized detection has been established for screening, diagnosis, treatment, and disease prevention for patients using digital data analytics. This revolution has not only aided clinicians, but remote diagnosis, faster consultation, door-step healthcare services are seeing a staggering rise in this sector through data analytics. Medical science has made unprecedented progress in the diagnosis and detection of infectious pandemics causing life-threatening situations. Every person is unique, and their response, as observed, has been different in dealing with COVID-19. It mainly includes the factors which can range from genetic make-up, their day-to-day lifestyle, and activities. Hence, pandemic detection and analysis through smart computing technologies depends on many factors. The factors include high-quality data collection, data linkage to large pools of other patient data through digital platforms.

Internet of things (IoT) and artificial intelligence (AI)-based algorithms have been so precise that based on individuals' unique characteristics, their risk of disease onset, type, and stage of disease infliction can be predicted [1]. The concept enables us to arrive at a deeper understanding of how to treat an individual. The digital analysis of patient data through smart computing technologies personalizes the diagnosis and enhances healthcare system for offering more precise treatment strategies. The individual's data generated drives the pharmaceutical industries to develop new medicines in specific ways towards more individualistic approaches. Hence, through the concept of precision medicine, patients are treated and diagnosed based on their personalized profile rather than treating everyone with the same amount or dose and type of therapy. It allows patients to live longer and have better lives by ensuring that every person is treated correctly. This is also giving rise to personalized healthcare technology tools. These tools aid physicians in assessing the likelihood of patients at risk of developing the disease and intervening to mitigate their impact. The smart systems can also wearables detecting physiological factors of the patients, or protein detection and DNA sequencing tools testing systems using machine learning (ML), deep learning, and AI to identify disease patterns and recognitions [2]. Using smart technologies in healthcare, especially if data is personalized, increases the likelihood of preventing disease development, early detection, individualized intervention for avoiding complications and adverse drug reactions. Therefore, medicine is now taking an approach towards collecting data of individuals, clinical trials involved, their genetic and environmental information to lay base models for algorithm expansion. Data analytics and smart technologies now work on patient-centric data to deliver personalized healthcare sustainable systems in the face of rising costs.

Today, numerous technological developments in electronics and information technology (IT) are evolving our modern life rapidly. These new technologies give academics, physicians, and patients a wide range of new opportunities and challenges. The new IoT and AI platforms are promising technologies in the health sector. IoT has been widely used to interconnect available medical devices and sensors that allow patients to monitor their medical conditions in real-time and doctors to manage their patients' health effectively and remotely. The ultimate objective of achieving high-quality practices depends on combining information received from heterogeneous sources efficiently and exchanging data while also protecting their security and confidentiality. Powerful data tools evaluate valuable information and have the potential of providing descriptive and personalized visualization of patients. Since, patients show varied symptoms, as clearly seen in COVID-19 patients, it becomes important to detect patients' risk of infection on an individual basis.

An innovative and tailored healthcare platform based on the IoT, the SM-IoT, and AI-based frameworks is used for COVID-19 patients and providers alike. Although digitalization impacts have been latent, remote patient connections have improved, and health services promoted smart computing technologies to lessen the burden of the healthcare economy affected by recent pandemic outbreaks in the last decade. SM-IoT can obtain information from diverse sources, incorporate it *via* a robust semantic web, store it in the cloud for further analysis, access this data by the user-friendly interfaces, and allow them to be accessed, protecting their privacy [3]. Healthcare data undergoes a considerable change. Changes vary from episodic to constant, illness to well-being, and from clinically regulated to patient inspired. Although there is already a capacity to create and record data, the innovations transform big data into intelligent data *via* qualitative and customized processing. Patients and healthcare practitioners may take worthier decisions and timely measures to promote enhanced personalized security. Since most efforts are focused to find a remedy or monitor COVID-19 spread have been ineffective, global surveillance of symptomatically and asymptomatically infected COVID-19 patients is challenging and becoming a demand. This will prepare the healthcare surveillance system better in case of lapse of future pandemics and create a multimodal data. Advanced embedded technology plays a significant role in multiple phases of different infectious diseases in the health sector, especially in the COVID-19 outbreak. The healthcare system has seen drawback in controlling the spread of infection as patient's profiles were varied. COVID-19 patients already diagnosed with non-communicable diseases (NCDs) such as cardiovascular diseases (CVD), diabetes mellitus (also known as type 2 diabetes-T2D), advanced respiratory distress syndrome (ARDS), cancer, etc., faced far more mortality rate than those not suffering from any complications [4]. Patient monitoring and diagnosis now require personalized and tailored treatment as the clinical response to COVID-19 attack was different in every individual. The only way to develop and incorporate new digital technologies at scale and speed for COVID-19 epidemic is to integrate it with a long line of modern public health data [5]. Digital technology can be integrated with online databases, data of illness cases and clusters, rapid tracing of contacts, tracking lockdown travel patterns, and communicating on a large scale. Hence, it can be accomplished through a more advanced digital and personalized healthcare system to tackle the outbreak of pandemics.

3.2   THE PERSONALIZED HEALTH CARE

Transition from evidentiary medicine is taking place. Large-scale clinical trials, meta-studies, and systemic analyzes define best practices for more customized medications that require individual-oriented diagnosis and care. Usual diagnosis and procedures focus on more granular groups with common characteristics or *people like me* rather than on a highly personalized treatment plan or medications that are tailored to fit human genotype and phenotype. The practice of medicine is, hence, changing. The processing and sharing of digital medical data provide new possibilities for gathering and using analytical methods from a broad range of sources. New digital platforms comparisons of clinical findings or cost-effectiveness analyzes for various drugs. With the ever-increasing accessible online information for patients, including medical records, clinical reports, diagnostics, current drugs, etc., the individual patient can now be more thoroughly understood as available medical information is growing every day. New methods and techniques for detecting diseases well before symptoms appear are developed for personalized care systems. The techniques include base risk assessments, new biomarkers, gene, and expressions and other enhanced systems for supporting the decision-making process (Figure 3.1) [6]. Ability to process vast quantities of data using analytical methods for clinical use ultimately depends on the kind of personalized care quality. A potential scenario can include clinical decision-making tools based upon new 'info-searching' or DeepQA technologies like Watson to help direct or advise doctors [7].

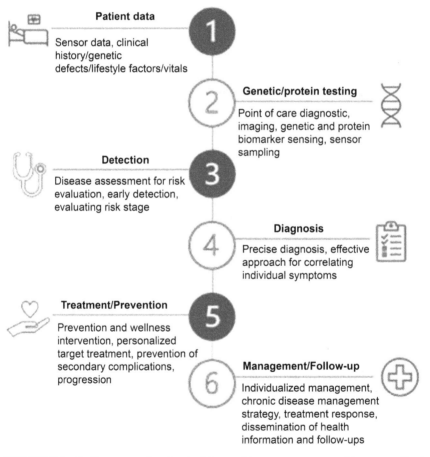

FIGURE 3.1   In-flow strategies for decision making process to provide personalized assessment and treatment for patients for prevention of chronic diseases and symptoms.

However, medicine is practiced in a health system, and personalized medicine is more than just practicing clinical medicine. To have more customized healthcare, we need to consider individual needs. Setting up an approach for extending the decision-making tools for COVID-19 detection as individuals tested positive for the novel coronavirus SARS-CoV-2 infection have shown unusual symptoms. This requires immediate planning towards personalized healthcare system set up for tracing, risk prediction, and therapy for individuals showing extreme response to pandemics. We also need to recognize and authenticate individuals correctly wherever they appear in the healthcare system and set up systems to protect the privacy and guarantee security.

The use of information and communication technology for clinical advantage has grown into a significant demand. Information infrastructure and technical programs for COVID-19 integrate data from treatment centers and occur in primary care physicians, community clinics, and hospitals. The control of the pandemic situations requires regions and countries to offer the e-Health infrastructure for better personalized care and purpose. COVID-19 detection using advanced tools and technologies has led to the development of various inventions and experiments using data analytics and smart technologies that have been established to communicate and trace individuals to provide individualized treatment across these platforms and facilities.

3.2.1 THE QUADRANTS OF PERSONALIZED HEALTHCARE SYSTEMS

The development of a personalized healthcare system is divided into different quadrants. The synchronization of a self-oriented system clinically directed approach, collaborative network, or collaborative/clinical guided system approach. The following examples further elaborate on this:

1. **Self-Oriented System:** This quadrant focuses on informing and training the individual. We are all aware of the internet as a common source of medical and health knowledge. The healthcare organizations and governments provide news and web pages to educate people about the risks, how people select their diets, symptoms, and medical procedures. In its survey of 'The Engaged ePatient Community' by the Pew Internet Foundation in 2008 [8], 80% of US

internet users searched for health information online, 75% indicated that their recent health searches influenced a decision about how to handle illness or disease. Thus, people are searching for and acting on the information. Although, it raises serious concerns about the need to determine the accuracy of health information on the Web.

2. **Clinically Directed Approach:** This quadrant is about tracking and educating the customer on programs that social services, hospitals, and third parties can provide [9]. It includes social warnings and alerts and remote control of patients. These programs can aim to foster the elderly's potential to live independently and enhance the quality of the care of those living with long and often self-managed conditions, including diabetes, congestive cardiac failure, COPD, and dementia [10]. Services may include sensors and panic buttons installed in the person's home, easily accessible medical devices such as scales and blood pressure cuffs and glucometers, and an app recording the patient's self-identified data such as mood, sleep habits, pain degree, and side effects on medicines. Being a promising field for research and investment, it may provide significant early indicators of decline or failure, thus reducing the need for admissions to emergencies. The pandemic surveillance services can also be included within this quadrant. Information from social media and online news sources can be combined with textual analysis to collect early warnings about infectious diseases and spread in the community.

3. **Collaborative System:** People with similar issues or circumstances are linked or related through 'Peer-to-Peer' health resources. It not only connects individuals locally or in a country, but globally. The individuals use these online platforms to locate, communicate, and exchange information with others with similar health issues or recent diagnoses. Providers may use the reviews given on these platforms to understand better and enhance the care processes. PatientsLikeMe is one of the primary sources of knowledge about life disorders such as multiple sclerosis and neuronal disease, where patients share their treatments, reports, simple health management system, and measurements [11].

4. **Collaborative/Clinical Guided System Approach:** This last quadrant focuses on decision-making and management needs to be considered as a core area for developing health co-creation. Doctors

and caregivers use online resources in this area to communicate with patients, reduce unnecessary face-to-face visits, and increase care continuity. Similar platforms are also developed especially for COVID-19 pandemic. The platforms that provide these skills are similar to those used on social media sites, such as e-mail, immediate message, and video, but are combined with medical recording systems and functional management systems to provide a robust platform. Other types of services in this quadrant include the 'shared care' model. It is typically based on a safe network that is shared between the patient and the healthcare team where goals can be identified, the outcomes stored, and the risk calculators are given to aid self-management based on the patient's profile [12]. The patient empowerment framework developed by IBM for Gachon University Gill in South Korea also offers patients online resources to better understand and manage their condition [13]. This is a typical example of tailored clinical guidance and information, real-time monitoring services, and effective collaborations towards personalization.

The growth of health systems is critical, especially after three pandemic outbreaks in the past decade. Medical practice is shifting, and a more personalized future is exciting. Simultaneously, our health systems are under tremendous pressure while maintaining the highest quality standards to become more effective, usable, and patient-oriented. Many residents and patients tend to be knowledgeable and inspire to take up a significant role in the co-construction of better health by using online resources and platforms to personalize their way of discovering, accessing, and maintaining communication with their jobs, families, and other social networks. Now is the time to develop and build a programed healthcare technology system that can provide personalized health care to everyone's good. Information resources are easily available as patients and their information are readily available, shareable, and approachable.

3.3 DIGITAL SURVEILLANCE OF COVID-19

3.3.1 DIGITIZATION OF COVID-19 DATA

A pandemic was announced by the World Health Organization (WHO) on 11 March 2020. In less than three months after the initial discovery of cases of COVID-19, a pre-known respiratory disease caused by the SARS-CoV-2 [14]. Similar to the preceding epidemic and pandemics, COVID-19 management focuses on the identification and containment of infection clusters, and the cessation of population spread to reduce the public health implications. The isolation of the affected areas and limitations on population movement can prevent further spread during the pestilence epidemic of the 14th century in Europe. Response to the public health outbreak initiatives is even essential today, including tracking, rapid case detection, community transmission interruption, and effective public communication. It is important to track the implementation and effects of these interventions on incidents and mortality.

The pandemic COVID-19 triggered the health and economic issues worldwide and led to one of the last century's biggest social crises. It is an indication of developments in modern healthcare technology, whether in front-line treatment, testing, or creative techniques. Developments that have been happening for tackling the COVID-19 crisis are in the field of service provision, AI, and data sharing, which can address the potential conclusions for detection, diagnosis, and prevention (Figure 3.2).

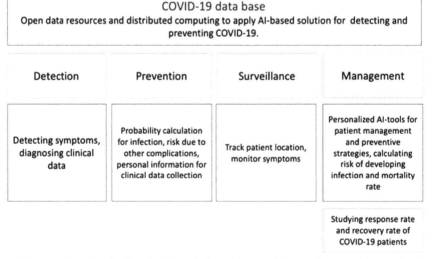

FIGURE 3.2   AI-application implementation at different stages of COVID-19 pandemic crisis.

Digital technology facilitates public health's global response to COVID-19, including population monitoring, case recognition, touch tracking, mobile data assessments, and public communication [15]. Thousands of cell phones, vast online data sets, related computers, relatively cheap computer resources, advancement in ML and natural language processing (NLP) take advantage of these rapid response procedures. This analysis seeks to capture the broad scope of digital technologies and their barriers, including legal, ethical, privacy challenges and organizations and employee barriers, to public health responses to COVID-19 worldwide. Public health is becoming more global as the need for international regulatory, evaluative, and emerging technology strategies integrate data to improve pandemic management and future planning for similar pandemic outbreaks like COVID-19. All countries have a core capacity to guarantee national preparedness for infectious risks, spread internationally, as mandated by International Health Regulations [16]. New and advanced methodologies and techniques for strengthening specific core capacities are currently being researched and built for a newer set of digital innovation. Hong Kong identified disease clusters using electronic data systems during the 2003 outbreak of SARS-CoV. Mobile data modeled travel patterns of patients during the Ebola outbreaks in West Africa between 2014 and 2016 [17]. Portable sequence devices permit improved understanding of the travel activities of epidemics and more efficient communication tracking. Digital technologies are deployed to reinforce the community spread of communicable diseases and pandemics. A system-level approach is essential to help prepare emerging solutions for potential epidemics and integrate them into COVID-19 management strategies. The epidemic management system created identifies the transmission of infections with respect to time, location, and individual to identify risk factors in the pandemic for successful intervention. Various digital data sources enhance and analyze the epidemiological of COVID-19 patients and their personalized data for integration into more external data sources obtained from hospitals and clinics.

3.3.2 ONLINE DATA SOURCES FOR EARLY COVID-19 DETECTION

Population monitoring technologies developed for COVID-19 currently rely on laboratory health data, clinician case reports, and syndromic surveillance networks (SSN). SSN tracking is focused on clinical symptom reports from a hospital, identified sentinel primary and secondary healthcare facilities, for example, 'influenza-like disease' rather than laboratory-based diagnostics, which promise to provide testing data for each case routinely. The detection of reported cases helps elucidate outbreak severity and properties and minimize further transmission. To fill the void with data from online news, news aggregation, social networks, web searches, and participatory longitudinal population cohorts, epidemiological data is gathered. The development of data aggregation systems for the online processing and filtering of data use NLP and ML algorithms. These data sources are gradually incorporated into the structured monitoring environment and monitor COVID-19 cases. The WHO EPI-BRAIN platform integrates numerous data sets, including environmental and meteorological data, for communicable diseases emergency preparedness and response [18]. Several systems reported the multi-purpose data and news reportage would identify timely disease detection for COVID-19 before the WHO stated the outbreak. The UK's automated syndrome monitoring system searches the National Health Service's digital records for breathing clusters that could signify COVID-19 [19]. Online data is used to estimate the actual spread of infectious diseases in the community. Based on proven influenza search algorithms on the internet, COVID-19 models are used as reports in public health areas making case studies more digitized and individual-centric. Syndromic monitoring, tracking an individual's response to the COVID-19 pandemic and necessary measures, is also enabled by crowdsourcing systems to elucidate the real disease burden. *InfluenzaNet* collects data from volunteers in many European countries on symptoms and social distance enforcement through a weekly survey [20]. In countries such as the United States of America, Canada, and Mexico, similar attempts occur, such as *COVID Near You* [21]. Although these systems are quick and insightful, they are subject to selection difficulties, over-interpretation of results, and failure to comply with official national supervision, which reports existing monitoring steps. A decentralized approach on these platforms means that individuals may further complicate the data without organized data collection and systematically performing developed algorithms for detection, diagnosis, and treatment (Figure 3.3).

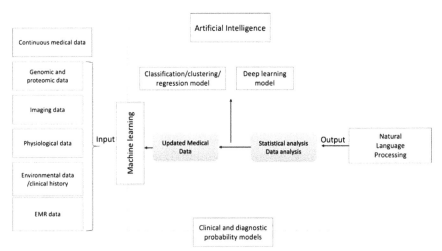

FIGURE 3.3 Predictive analytics of healthcare data for personalizing diagnosis and detection system using algorithm models.

3.3.3   DATA-VISUALIZATION TOOLS FOR COVID-19 DECISION SUPPORT

Data visualization tools keep individuals updated and assist policymakers in refinement measures. The data dashboards are commonly utilized throughout the epidemic and capture real-time data on public health, along with reported incidents, fatalities, and testing with intense patient tracking and individual monitoring. COVID-19 control panels usually concentrate on time series tables, spatial maps, therefore coordinating data from regional statistics to cases. Several dashboards show broader pandemic reactions, for example, clinical trials, strategic, and economic measures of policymaking, and actions against for incorporating socially distant guidelines. Few dashboards include data apps for contact tracking and group monitoring. There are also questions about obstacles to the accuracy and integrity of data collection. Lack of government-wide data detailing official requirements and discrepancies in countries complicates global comparisons. Statements are often not always open to updating and also not reliable for observing statistics. New visualization methods have overcome these obstacles; for example, the NextStrain Open repository builds an overall map of infection spread [22]. It allows open data sharing and open-source technology. In previous global outbreaks, such pace of sharing data was not observed. However, COVID-19 tracking has made early detection easy.

3.3.4   RAPID AND INDIVIDUALIZED CASE IDENTIFICATION

Prompt and timely case detection is vital during an epidemic to minimize the further spread of the critical risks and transmission modes. Digitization may improve clinical and laboratory communication by using case signs and broad access to population research and self-mode testing. Digitization automats and speeds up public-health reporting. Online symptom reporting of individual case detection is generally used for tracking but provides advice on isolation and access to other health facilities, including video assessments and testing. Similar programs should be introduced quickly but also linked to active monitoring and intervention in the field of public health, such as case isolation and touch quarantines. Although this method serves symptomatic patients, extensive tracking and touch recording of individuals and communities and their health stats with their disease history can play a critical and fundamental role in detecting up to 80% of COVID-19 cases being moderate to asymptomatic. To distinguish possible cases based on febrile symptoms (for example, at airports), thermal sensors or imaging cameras are positioned. A large number of false-positive or -negative findings indicate that, beyond rising knowledge, this is unlikely to have a significant effect. The tracking of COVID-19 in populations is also explored in wearable technologies and tracking an individual's vital stats. It also provides health data for future reference in case of other pandemic outbursts, protecting the vulnerable populations. It lowers the risks of other infections, especially in those whose immunity weakens over time or weakened due to COVID-19.

Digitally connected and fast diagnostic tests involve broadened access to research, capacity building, and healthcare systems and diagnostic labs. Several COVID-19 PCR experiments are currently carried out. They are primarily used in healthcare settings. Access to tests has improved *via* testing installations and automated swab kits. Sampling, shipping of samples to centralized laboratories, anticipating for results, and follow-ups are inherent delays. In comparison, rapid diagnostic antibody testing in the home, community, or social care environment could be carried out and yield results in a matter of minutes. The use of the picture processing and ML methods to link smartphones to automated re-reading can easily connect the mass tests with geospatial and patient data, both in clinical and public health systems, to speed up the results. Standardization and incorporation of electronic data of patient records are essential for this to work effectively. Another set of technical advancement is also offering computerized topographical scans of patient screening through ML algorithms. Hence, rapid tracking and quarantining of contacts is required after case identification and isolation to prevent additional transmission. In areas with high communications, these steps are enforced and controlled on an increasingly inaccessible or at least historically challenging scale.

3.3.5   EVALUATING PERSONALIZED INTERVENTIONS THROUGH THE USE OF DATA MOBILITY

Smartphone aggregate individuals' location-specific data through GPS, mobile networks, and Wi-Fi. These platforms can track population flows in real-time, identify possible hotspots of transmission and provide insights into the efficacy of health-related measures, such as travel limitations for observing definite human behavior. Access to mobility data is a significant issue, however, questions have been raised about ethics and privacy which use these methods and are not aided in providing treatment but can prevent further transmission. Recently, for COVID-19 control purposes, many technologies and telecommunication companies have made available mobility data with privacy-preserving aggregation steps; the data sets are nevertheless restricted, and a long-term data-sharing agreement created. Mobile information and sensing technologies will reduce health care costs and enhance health research and performance. These innovations will promote continued human and population-based health surveillance and facilitate healthier habits to avoid or minimize health conditions. Better self-management of chronic illnesses, clinicians' awareness, reduction in the number of health visits and deliver previously impossible, personalized, accessible, and on-demand services can be a reality. The innovations track health status and enhance health outcomes and aid clinical decisions. The digital technologies have been developed and tested for COVID-19 patients, especially for those who have previous diseases such as diabetes, asthma, obesity, smoking cessation, stress control, and depression, and are at much greater risk of getting infected by SARS-CoV-2. However, it is still unclear if e-Health tools can contribute to better overall health results and lower disease burdens. For example, in recent research, it is noted that SMS (short messaging services) based health interventions are not properly assessed for their effectiveness. There is little evidence that mobile applications have been evaluated for smoking cessation. mHealth equipment, apps, and services cannot function or could, at worst, have harmful effects on health and its related cost compromising quality. Untested e-Health innovations may be prematurely implemented in a health structure already burdened with not so optimal results and delivering higher costs than expected. E-Health innovations promote new approaches for the processing effects of biological, compartmental, or environmental data. Sensors that track phenomena with greater accuracy increased sampling frequency, less data loss, better, and improved comfort, and, in some cases, reduced costs as compared to conventional steps. Sensor data and self-reporting algorithms permit physiological, social, responsive, and environmental inferences, for instance, psychological stress systems, smartphone sensors, or smoking.

3.3.6   MODEL-BASED DESIGN OF ADAPTIVE AND PERSONALIZED MEDIATIONS USING SMART COMPUTING TECHNOLOGIES

Smart computing technologies carry the potential to individualize and personalize interventions in real-time. This leads to incorporation of interventions that minimize waste, augments compliance by increase the power of intervention. In order to measure the internal difference between patients and their response to COVID-19, mediating variables of every individual are recorded digitally as clinical patient history. More statistical methods need to be introduced which specify the effects of COVID-19 between and within patients to target better-personalized intervention. For example, it may be necessary for tailoring and personalizing approaches to know if shifts in mood contribute to healthy habits such as smoking, eating, or exercise. The healthcare crisis COVID-19 has caused, the recent development of tailoring interventions has begun by screening either factorial or fractional variance design analysis. Hence, statistical enhancements and modeling the patient and target treatment through ML enable the speed and evaluation of personalizing treatment against traditional approaches.

3.4 HIGH DATA COLLECTION, PROCESSING, AND ANALYTICS FOR COVID-19

3.4.1 HIGH-DENSITY DATA COLLECTION

High-end clinical data of COVID-19 patients is available for research opportunities and provide monitoring and intervention in real-time. The healthcare research with the raw data of patients can give scalability as it can help prepare the healthcare services for future pandemics. The pool of data now needs digitization and makes it more efficient for providing personalized therapy.

Digitization can provide high sampling rate and support the quantification and classification of COVID-19 data. The high-density data available can amplify the degree and discriminative power of the experimental design under study. Raw data compiled is the new "fingerprint" of every individual. The status of disease response has been so varied that a single pill concept has not worked. It also calls for a better explanation as to why personalization is required. Such intensive longitudinal and high-density unique data for every individual allows the need to examine the effects of variance, extra variable factors, close-connection between or within the individuals, and incorporate parameter of interest in digitizing the data.

3.4.2 DATA PROCESSING

High-density data requires precise processing, especially for clinical studies, due to extreme inpatient data variability. ML models and methods make processing and classification decisions by selecting features from the patient data and segment it to draw inferences about individuals. COVID-19 data of every individual processed for accurate analytics and high-frequency data can provide personalized intervention. Data classification methods include unsupervised cluster analysis, latent class analysis, topic models for learning association among data, and network patterns. Data can be presented in numeric or textual data. If data for training is accessible, unsupervised clustering is more efficient in data coalition.

Another popular technique for classification is vector machines (SVM) [23]. The technique can be used to robust performance in classification; however, sophisticated algorithms, for instance, decision-trees can give adequate clinical data accuracy. Variables of interest cannot be complied explicitly within some instances and must be extracted from variables that can be calculated directly. In such cases, the inferences should be calculated using different models, for example, hidden Markov models [23]. Similarly, the factor analysis and latent characteristic models may show the internal data structure so that the discrepancy in calculated variables or elements is better clarified. These data classification processes and inference sets a minimized bias to achieve the high precision of a data classifier for data, which varies enormously, such as clinical data among COVID-19 individuals. The multi-step process that predicts the outcome or estimates possible outcome or therapy for an individual (1) classifies the data; (2) stratifies the annotated data for training and validation; (3) determines the type of machine-learning classifiers for testing; (4) sets time segments or data window of the data streamed to apply the appropriate classifiers.; (5) extracts data features relevant as an input to the ML classifier (mean, variance or amplitude, degree, etc.); and (6) tests the classifier accuracy agreement with the trained dataset then with the data to be validated. This way, the algorithms are trained for an output. These steps are also carried out in an iterative process to evaluate the best research technique for data through different classifiers, window sizes, and data characteristics.

3.4.3   REAL-TIME DATA ANALYSIS

Many data measures obtained from COVID-19 in real-time require subject participation and use data for high-end processing collected from hospitals, communities, surveys, and locations. It enables studies to execute real-time data analysis for a more precise outcome and control various therapeutic and experimental approaches on patients and design human clinical trials. The models using real-time data analysis can reduce the time and execute more precise clinical predictive models' detection, diagnosis, and therapy. The predictive modeling can properly select evidence-based data and help design a more personalized COVID-19 patients' diagnosis. Real-time data analysis requires gathering data from multiple sources such as physiological, behavioral, environmental, and biological data of COVID-19 patients. Surveys can also help collect first-line data for analysis and open more factors to be integrated with algorithms. These algorithms can be put into clinical utility and explore linkages between all data types for more comprehensive diagnosis and therapeutic decisions for patients. Combing and data fusing for analysis are based on the principle of permutation and combination to produce probabilistic techniques allowing researchers and clinicians to interpret and assess the data based on the available decision's healthcare workers make. The high-end smart analysis may even bring out the validity and reliability of the treatment offered to patients. The data variability can be reduced using the high statistical power and fuse data for more segmentation. All measurements cannot be made at the same spatial and time resolution is difficult for fusion in practice. The multiscale and multimodal information can help curb challenges in research for future pandemics and prepare clinicians to make a more immediate and precise decision for therapy.

3.4.4 INTERVENTION OUTCOME ASSESSMENT

Real-time monitoring of data analytics assesses patients' intervention outcomes and how the pandemic has influenced adherence behaviors. Intervention outcomes can be linked to individual data, automate the detection of a patient's condition, infer the intervention, and monitor the intervention outcome. The effects of intervention dissemination are essential because it can serve as a base model to apprehend the impact of the intervention provided to the patients tested positive for COVID-19 and further add the outcome data to the algorithms. Evidence-based medicine can offer new data digitization capabilities and lead to better evidence of providing interventions and study outcomes for the quick clinical decision-making process.

3.5 ARTIFICIAL INTELLIGENCE (AI) AND SARS-COV-2

The COVID-19 pandemic due to the novel SARS-CoV-2 is ongoing, and the importance of emerging technology for the pandemic response is too early to be completely quantified. Although digital technologies offer pandemic response tools, they are not the complete answer. The consensus emerges that the outbreaks and pandemics play a significant role in supplementing traditional public-health interventions, thus ease the human and economic effects due to COVID-19. Rapid and wide-spread research of automated symptomatic controls, epidemiological expertise, and long-term clinical follow-up requires cost-effectivity and sustainability systems-level approaches to develop modern early detection systems and personalized diagnoses. Public health will eventually become more and more global, and the role of emerging technologies has become critical in pandemic preparedness plans. Key digital players, including technology firms, should be a preparation for long-term partners rather than during continued emergencies. Viruses are infinite and digital technology, and data are, increasingly, limitless. International methods for monitoring, analyzing, and using emerging technologies need coordination to improve the pandemic and plans for the COVID-19 outbreak and other infectious diseases. Primary care and outpatient hospital services have been exciting for some time, but the extent of digital transformation has been minimal until quite recently. In the face of many challenges to repayment, accreditation, and human conditions, digitization, and related expectations need to be balanced for the system to adopt much better strategies. In a matter of weeks due to COVID-19 pandemic, the scenario has changed drastically. Many countries have implemented first-class digital programs, remote monitoring, and television medical services without physical contact. Furthermore, it has been well recognized and catalyzed that digital health technology can diagnose patients and provide fast service to the detection system. Three reasons made this dramatic change possible.

For example, many businesses may provide solutions, instead of starting from scratch, by modifying existing software. When COVID-19 struck, the technology was sufficiently mature to be used in the scale. For many patients in primary and outpatient clinics, remote management is possible. The critical question is whether healthcare services can, after the COVID-19 pandemic, return to interactions more face-to-face. Some digital remote technologies, such as digital primary treatment, have not been explored to date. In the vaccine and treatment process, AI has also been used. AI can help health care systems tackle COVID-19 by anticipating, minimizing, and promoting diagnosis and treatment. Imaging diagnostics can further contribute to the powerful impact on the use of health data and disease monitoring. There are unanswered concerns about data accuracy, data transferability across environments and healthcare systems, algorithms efficiency when used in clinical procedures, and data access and privacy protection. The crisis gives us the chance to gain insight into the future and to think about those problems.

3.5.1 DATA RECONSTRUCTION USING AI

Prior to the COVID-19 worldwide pandemic penetrated the healthcare system, there were high hopes that the widespread development can be delayed. AI-based digital platforms within medical care enable over-extended healthcare providers to develop new drugs, optimize data and information flow, and to provide personalized and timely treatment. AI-based study of social media and coverages predict the outbreak's emergence at the beginning of the current crisis. A Canadian-based company, Blue Dot, is credited with identifying the rare group of pneumonia-like cases in Wuhan, and official sources confirmed the results as COVID-19 infection [24]. Great amounts of data can be congregated and analyzed quickly from a range of sources to reconstruct and predict the potential spread and the behavior of the COVID-19 outburst. Early promising models, such as these, and predictive behavior, can question data's reliability and accuracy that go into the AI-based analysis must be challenged when providing therapeutic models for patients. Patient diagnostics and personalized therapy using clinical data can be expanded for the symptom management framework with a specific algorithm for the decision of COVID-19 that could help patients develop and get more tailored advice. This could potentially decrease the number of emergency rooms and walk-in facilities unnecessary. But no conclusive proof is available at this time.

3.5.2 IMAGING DIAGNOSTICS USING AI FOR COVID-19

Imaging and diagnostics application area of AI during the pandemic may hold the key to strengthen the loopholes in controlling pandemics. If successfully implemented in clinical practice, the AI approach can improve the clinical decision-making process and potentially monitor patients' phase. Medical-imaging-based methods for COVID-19 have been implemented, especially in CT scans [25]. CT-based COVID-19 diagnosis is a precise and sensitive approach to view structural inflammation or damage to internal organs of the patients infected with SARS-CoV-2.

To speed up and boost their performance, the shared data repositories developed globally, report with much improved clarity. The data repository accelerates the development of algorithms and enhance their performance.

Through imaging and diagnostics, AI helps COVID-19 therapy design new pharmaceutical products and redeploy existing medicines. For example, several research papers available as open research database in relation to COVID-19. The data can be analyzed using ML, so the relevant information on medicines can support the interventions against COVID-19. AI accelerates the diagnosis for COVID-19, and open data initiatives. AI-based CT scans are the first imaging used for COVID-19 diagnosis, and the data is now linked to individuals' signs and severity of symptoms. The inference from AI adds data to a personalized diagnosis and considers the therapy level to be offered to the patient. Research groups are using a deep-learning approach to build AI algorithms for chest CT-scan. The sensitivity achieved using the prototype is 98.2% [26] and 92.2% sensitivity for the thoracic CT algorithm. AI-based algorithms in medical imaging focus on the segmentation of images. The regions identified and quantified for damage account for the severity of the lung's chronic damage due to SARS-CoV-2 infection.

3.6 AUGMENTED PERSONALIZED HEALTH DATA USING IOTS AND AI

The customized digital health data renders ample information through contextual and personalized processing. It covers disease-, lifestyle-, clinical-, cyber-social-centered multidimensional data to implement personalization. This data will bring about informed choices after being translated to customizing digital health programming using AI.

3.6.1 AUGMENTED PERSONALIZED HEALTHCARE (APH)

Health care is expected to change by using the required wearables, IoT, mobile apps, e-medical registers (EMRs), and social media to personalize all related physical, cyber, and social data [27–29]. The data has to expand integrating all signals and information and adding medical expertise and AI techniques for a better outcome, as shown in Figure 3.3. Sensor data of the populations can directly translate these signals and measure patients' health outcomes. COVID-19 patient data has to be under constant supervision for collecting relevant information for calls to be generated at the patient, clinician, and demographic levels. APH includes all emphasis on patients and the diseases for further infection spread. This will also provide disease prevention measures, future prediction of conditions, and information on harmful side effects of the infection. Critical enablers for APH using IoT and AI and providing progressive ability to capture, evaluate, and manipulate big data for a COVID-19 individual as follows:

- Cost-effective sensors to build a multimodal algorithm using their personalized data of physiological, biological, and lifestyle data (including diseases history, meals, and recovery from or during COVID-19).
- Keep access or update patient clinical records and follow-ups. Connecting the data to online repositories and peer-reviewed medical literature online (PubMed).

Highly personalized healthcare should be able to perform analysis on extreme variable responses from patients and their conditions due to COVID-19. Available drugs recommended to infected patients showed an adverse reaction in a few while others were able to recover, and some ended up recovering but showed various long-term side effects. Personalized healthcare can provide the right dose, amount, and type of drug to patients for their recovery, depending on their condition severity and trigger prevalence. Pandemic control, therefore, calls for a more thorough evaluation, especially analyzing data of cause and vulnerabilities. Varied data can always give successful decision-making aid to clinicians.

3.6.2   COGNITIVE COMPUTING (CC) FOR ADVANCED PERSONALIZED HEALTHCARE

Cognitive computing (CC) builds multimodal data in response to the interpretation of patient data [30]. It considers clinical history, lifestyle, and physical features, environmental variables, age, weight, and patients' health activities. The generation of hybrid techniques combines probabilistic and declarative models to scale up individual diagnoses. ML and AI-based techniques discover associations of the data's predictive capacity and the influence of variables, which is case to case. This way, clinical data can advance towards personalized health care. With frequent encounters of three pandemics in the last decade-SARS-CoV, MERS, and SRAS-CoV-2 require faster diagnosis and preventive measures to avoid documentation, and future pandemics are likely. CC can leverage personalization and aid clinicians for better diagnosis and implement individual-based response for therapy. CC builds multimodal data frameworks utilizing self-learning skills built on feature selection, data mining, pattern recognition, natural language, and human sense processing and creates automated models on the clinical data. It segments healthcare data and expands the NLP frameworks to provide clinicians with decision-making tools and personalized treatment modules. It advances preciseness to healthcare services. Till now, healthcare data was unstructured. CC is utilizing healthcare records at hospitals, published clinical and trail data to assist diagnosis more significant. The cognitive analytics on healthcare information can infer the risk of patients encountering a particular disease or condition.

3.6.3   PERCEPTUAL COMPUTING (PC) FOR ADVANCED PERSONALIZED HEALTHCARE

Perceptual computing (PC) focuses on domain rich information linking causes to effects and reasoning techniques [31]. The prediction of the impact of patient data, PC interprets and evaluates the unclear or missing data. The model is constructed based on incorporating data automatically in a precise way to eliminate the uncertainty. Patient data can vary. It can change during the recording or methodology of data grouping. It is challenging to construct appropriate data abstracts and organize segments of grouped data in a homogenous manner. In general, patient symptoms are based on their characteristics, vulnerability/susceptibility, preventative measures, patient medication and procedures, and the severity of causes. One of the most open-ended problems is how to synthesize a patient's vulnerability report in support of a transparent health care purpose. Doctors occasionally monitor chronic diseases at defined intervals. Information to the doctor is available through verbal communication of the patient perspective (self-reporting); unfortunately, the critical information is neither recalled by the patient nor reported to the doctors. Hence, it requires an adequately defined track that impacts the problems above, to measure and convey remedial measures in ways that can be readily accessible to end-users (patients or clinicians). The nature of chronic conditions and lay users (patients) accessibility is complicated. Perception analysis utilizes somaticized and **CC** to optimize clinical outcomes for the patient and clinician for timely, adaptive measures. The data needs to be turned into useful information for clinical decision-making aid.

3.7 DIGITAL HEALTHCARE FOR DETECTION OF FUTURE PANDEMICS

Pandemics such as corona outbreaks need practical implementations for diagnostics and therapeutic purposes. Digital platforms for data coalition are a must when the response to such epidemics is robust and varied among individuals. Key challenges arise when the data has to be personalized to avoid potentially harmful and adverse drug reactions. Digital medicine provides a route to accessibility and affordability of quality care, especially where expert and analytical advice are limited. IoT and AI can prevent the spread of diseases, help diagnose the disease stage, and provide a timely decision to individuals in a more tailored approach.

3.7.1 DIGITAL HEALTH STRATEGY

Traditional healthcare practices should not be replaced but should be improvised for speed and precision decision-making, so trial and error can be avoided. Digital healthcare can: (i) monitor patients' health more closely; (ii) diagnose health more accurately; (iii) develop personalized prevention and treatment plans; (iv) provide access to wider audiences; (v) improve the quality of patient life. Organizations and industries have to create more strategies to implement and segment data to develop advanced algorithms for providing personalized healthcare and interventions for future pandemics. Real-time and data-driven IoT and AI technologies can construct high-throughput information to a single platform and add new data compiled over the years to provide immediate/short-term initiatives and long-term intervention.

3.7.2 PRECISION MEDICINE AND ARTIFICIAL INTELLIGENCE (AI)

Precision healthcare and preventive approach lead to precision medicine. Dynamic algorithms developed through IoT and AI-based technologies can bring post-COVID times into clinical practice (Figure 3.4). The comprehensive healthcare system reforms, high-performance computing (HPC), biological datasets, and implements precision medicine pathways with multidimensional data curated from refining unstructured data. A personalized healthcare system requires precision medicine to predict potential therapy further and optimize based on pattern recognition.

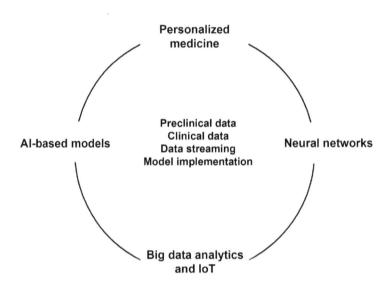

FIGURE 3.4 The correlation of emerging technologies to implement personalized medicine.

Methods for precision medicine include neural network, support vector, random forest, and evolutionary algorithms (EA) [32]. As COVID-19, patients experience co-morbid conditions (multiple clinical complications), especially in patients already suffering from NCDs or the elderly populations with the highest mortality rate, predicting treatment responsiveness in the vulnerable population through genetic and phenotypic becomes essential. Integrating the OMICS data within the AI-drug development algorithm as the next step to personalized healthcare will enhance precision medicine for COVID-19 therapy. AI algorithms in longitudinal studies can help capture development patterns and variability over time at a level of the individual or subpopulation.

3.8   LEGAL, ETHICAL, AND PRIVACY CONCERNS IN IOT AND AI-BASED DIGITAL PLATFORMS

COVID-19 pandemic has garnered robust data of an individual's response to the infection. The epidemic has proved that an individual's response is varied based on their biological, physiological, genetic, environmental, and lifestyle history. The patients' data collection holds legal issues, ethical concerns for using their data for developing digital healthcare platforms [33]. Privacy concerns pose deep ethical problems as personal data is used for continuous surveillance. Sensitive information still needs to undergo proper efficacy, which underlines the need for patient faith and participation with full consent. All systems create a digital platform, mostly a personalized platform involving and collecting multimodal patient data and conducting follow-ups. It also includes tracking their locations, and it must comply with the requisite legal, ethical, and clinical governance. All data-driven from patients should go under a legal contract. Independent auditing should be held to ensure that data is not used for other external use other than analyzing the pandemic course. The dynamic algorithms should be subjected to cybersecurity rules and regulations and have tamperproof encryptions to keep data secured [34].

3.9   FUTURE DIRECTIONS AND CONCLUSION

The future of innovating IoT and AI-based digital and personalized healthcare system for pandemic outbreaks have adopted more excellent computing capabilities. The advanced algorithms used for detection, prevention, and therapeutic purpose will exploit computational efficiency to create a massive digital platform and open new cases and information. Digital platforms need to achieve more incredible speed and ability to handle large data sets. The personalized healthcare system with high computational algorithms will change how the healthcare system perceives biology and its essential clinical utility. Every individual will be served more precisely, and AI-based tools and technologies will advance personalized therapy towards better decision making and centralize the diagnosis, prognosis, and treatment on every individual and their personal data combined. This will bring out more reliable predictions and follow-ups, especially for vulnerable populations, and prevent the spread of similar infections. Simple input and output data processing methods have been the focus of AI, ML, and statistical analysis research. The advanced platforms to fight future pandemics will be more individual-centric and identify which intervention might provide the right action mechanism. The application of IoT and AI is bringing sensor-based data into focus and insight into continuous disease monitoring to make data more personalized and precise. This reflects those days are not far where patients will be timely predicted for the risk of developing a particular disease or infection. COVID-19 has given it yet another push towards bringing the healthcare industry to focus on creating digital and personalized healthcare system to prepare us better to tackle a near future pandemic.

KEYWORDS

- **artificial intelligence**
- **cardiovascular diseases**
- **coronavirus diseases**
- **information technology**
- **internet of things**
- **middle east respiratory syndrome**
- **non-communicable diseases**
- **severe acute respiratory syndrome-corona virus**

REFERENCES

1. Talukder, A. K., Chaitanya, M., Arnold, D., & Sakurai, K., (2018). Proof of disease: A blockchain consensus protocol for accurate medical decisions and reducing the disease burden. In: *2018 IEEE Smart World, Ubiquitous Intelligence & Computing, Advanced & Trusted Computing, Scalable Computing & Communications, Cloud & Big Data Computing, Internet of People and Smart City Innovation (SmartWorld/SCALCOM/UIC/ATC/CBDCom/IOP/SCI)* (pp. 257–262). doi: 10.1109/SmartWorld.2018.00079.

2. Topol, E. J., (2019). High-performance medicine: The convergence of human and artificial intelligence. *Nature Medicine, 25*(1), 44–56. Nature Publishing Group. doi: 10.1038/s41591-018-0300-7.

3. Dridi, A., Sassi, S., & Faiz, S., (2017). Towards a semantic medical internet of things. In: *Proceedings of IEEE/ACS International Conference on Computer Systems and Applications, AICCSA*, 1421–1428. doi: 10.1109/AICCSA.2017.194.

4. Azarpazhooh, M. R., et al., (2020). COVID-19 pandemic and burden of non-communicable diseases: An ecological study on data of 185 countries. *Journal of Stroke and Cerebrovascular Diseases, 29*(9), 105089. doi: 10.1016/j.jstrokecerebrovasdis.2020.105089.

5. Holmes, E. A., et al., (2020). Multidisciplinary research priorities for the COVID-19 pandemic: A call for action for mental health science. *The Lancet Psychiatry, 7*(6), 547–560. Elsevier Ltd. doi: 10.1016/S2215-0366(20)30168-1.

6. Lesko, L. J., & Atkinson, J., (2001). Use of biomarkers and surrogate endpoints in drug development and regulatory decision making: Criteria, validation, strategies. *Annual Review of Pharmacology and Toxicology, 41*. 347–366. doi: 10.1146/annurev.pharmtox.41.1.347.

7. Ferrucci, D., Levas, A., Bagchi, S., Gondek, D., & Mueller, E. T., (2013). Watson: Beyond Jeopardy! *Artificial Intelligence, 199, 200*, 93–105. Elsevier B.V. doi: 10.1016/j.artint.2012.06.009.

8. *The Engaged E-patient Population* | Pew Research Center. https://www.pewresearch.org/internet/2008/08/26/the-engaged-e-patient-population/ (accessed on 22 June 2021).

9. Klassen, A., Miller, A., Anderson, N., Shen, J., Schiariti, V., & O'Donnell, M., (2009). Performance measurement and improvement frameworks in health, education and social services systems: A systematic review. *International Journal for Quality in Health Care, 22*(1), 44–69. Oxford University Press. doi: 10.1093/intqhc/mzp057.

10. Mukamel, D. B., et al., (2006). Team performance and risk-adjusted health outcomes in the program of all-inclusive care for the elderly (PACE). *Gerontologist, 46*(2), 227–237. doi: 10.1093/geront/46.2.227.

11. McCaffrey, S. A., Chiauzzi, E., Chan, C., & Hoole, M., (2019). Understanding 'good health care' from the patient's perspective: Development of a conceptual model using group concept mapping. *Patient, 12*(1), 83–95. doi: 10.1007/s40271-018-0320-x.

12. Roblin, D. W., (2011). The potential of cellular technology to mediate social networks for support of chronic disease self-management. *Journal of Health Communication, 16*(1), 59–76. doi: 10.1080/10810730.2011.596610.

13. IBM, (2019). IBM Watson health in oncology. *Scientific Evidence*, pp 1–33. Available at: https://www.ibm.com/downloads/cas/NPDPLDEZ (accessed on 14 July 2021).

14. Sharma, A., Tiwari, S., Deb, M. K., & Marty, J. L., (2020). Severe acute respiratory syndrome coronavirus-2 (SARS-CoV-2): A global pandemic and treatment strategies. *International Journal of Antimicrobial Agents, 56*(2), 106054. doi: 10.1016/j.ijantimicag.2020.106054.

15. Whaiduzzaman, M., et al., (2020). A privacy-preserving mobile and fog computing framework to trace and prevent COVID-19 community transmission. *IEEE Journal of Biomedical and Health Informatics* (Vol. 24, No. 12, pp. 3564–3575). doi: 10.1109/JBHI.2020.3026060.

16. WHO, (2017). *International Health Regulations (2005): Areas of Work for Implementation* (pp. 1–32). Available at: https://www.who.int/ihr/finalversion9Nov07.pdf (accessed on 14 July 2021).

17. Cohen, N. J., et al., (2016). Travel and border health measures to prevent the international spread of Ebola. *MMWR Supplements, 65*(3), 57–67. doi: 10.15585/mmwr.su6503a9.

18. Schafer, I. J., Knudsen, E., McNamara, L. A., Agnihotri, S., Rollin, P. E., & Islam, A., (2016). The epi info viral hemorrhagic fever (VHF) application: A resource for outbreak data management and contact tracing in the 2014–2016 west Africa Ebola epidemic. In: *Journal of Infectious Diseases* (Vol. 214, No. 3, pp. S122–S136). doi: 10.1093/infdis/jiw272.

19. Cheng, M. P., et al., (2020). Diagnostic testing for severe acute respiratory syndrome-related coronavirus 2: A narrative review. *Annals of Internal Medicine, 172*(11), 726–734. NLM (Medline). doi: 10.7326/M20-1301.

20. Koppeschaar, C. E., et al., (2017). Influenzanet: Citizens among 10 countries collaborating to Monitor influenza in Europe. *JMIR Public Health and Surveillance, 3*(3), e66. doi: 10.2196/publichealth.7429.

21. Baumgart, D. C., (2020). Digital advantage in the COVID-19 response: Perspective from Canada's largest integrated digitalized healthcare system. *NPJ Digital Medicine, 3*(1), 1–4. Nature Research. doi: 10.1038/s41746-020-00326-y.

22. Hadfield, J., et al., (2019). Twenty years of West Nile virus spread and evolution in the Americas visualized by next strain. *PLoS Pathogens, 15*(10), e1008042. Public Library of Science. doi: 10.1371/journal.ppat.1008042.

23. Miao, Q., Huang, H. Z., & Fan, X., (2007). A comparison study of support vector machines and hidden Markov models in machinery condition monitoring. *Journal of Mechanical Science and Technology, 21*(4), 607–615. doi: 10.1007/BF03026965.

24. Kumaravel, S. K., et al., (2020). Investigation on the impacts of COVID-19 quarantine on society and environment: Preventive measures and supportive technologies. *3 Biotech, 10*(9), 393. Springer. doi: 10.1007/s13205-020-02382-3.

25. Zhu, Y., et al., (2020). Clinical and CT imaging features of 2019 novel coronavirus disease (COVID-19). *The Journal of Infection, 81*(1), 147. NLM (Medline). doi: 10.1016/j.jinf.2020.02.022.

26. Gozes, O., et al., (2020). *Rapid AI Development Cycle for the Coronavirus (COVID-19) Pandemic: Initial Results for Automated Detection & Patient Monitoring Using Deep Learning CT Image Analysis.* [Online]. Available: http://arxiv.org/abs/2003.05037 (accessed on 22 June 2021).

27. Behera, R. K., Bala, P. K., & Dhir, A., (2019). The emerging role of cognitive computing in healthcare: A systematic literature review. *International Journal of Medical Informatics, 129*, 154–166. Elsevier Ireland Ltd. doi: 10.1016/j.ijmedinf.2019.04.024.

28. Bhalla, N., Pan, Y., Yang, Z., & Payam, A. F., (2020). Opportunities and challenges for biosensors and nanoscale analytical tools for pandemics: COVID-19. *ACS Nano, 14*(7), 7783–7807. American Chemical Society. doi: 10.1021/acsnano.0c04421.

29. Sheth, A., Jaimini, U., & Yip, H. Y., (2018). How will the internet of things enable augmented personalized health? *IEEE Intelligent Systems, 33*(1), 89–97. doi: 10.1109/MIS.2018.012001556.

30. Sheth, A., Jaimini, U., Thirunarayan, K., & Banerjee, T., (2017). *Augmented Personalized Health: How Smart Data with IoTs and AI is about to Change Healthcare.* doi: 10.1109/RTSI.2017.8065963.

31. Madnick, S. E., Wang, R. Y., Lee, Y. W., & Zhu, H., (2009). Overview and framework for Data and information quality research. *Journal of Data and Information Quality, 1*(1), 1–22. doi: 10.1145/1515693.1516680.

32. Nezhad, M. Z., Zhu, D., Li, X., Yang, K., & Levy, P., (2017). SAFS: A deep feature selection approach for precision medicine. In: *Proceedings - 2016 IEEE International Conference on Bioinformatics and Biomedicine, BIBM 2016* (pp. 501–506). doi: 10.1109/BIBM.2016.7822569.

33. Hathaliya, J. J., & Tanwar, S., (2020). An exhaustive survey on security and privacy issues in healthcare 4.0. *Computer Communications, 153,* 311–335. Elsevier B.V. doi: 10.1016/j.comcom.2020.02.018.

34. Cardenas, A. A., Amin, S., Sinopoli, B., Giani, A., Perrig, A., & Sastry, S., (2009). *Challenges for Securing Cyber-Physical Systems 5*(1).

CHAPTER 4

Impact of Lockdown on Social and Mobile Networks During the COVID-19 Epidemic: A Case Study of Uttarakhand

PRACHI JOSHI[1] and BHAGWATI PRASAD PANDE[2]

[1]*Assistant Professor, Department of Information Technology, Almora Campus, Soban Singh Jeena University, Almora, Uttarakhand, India, E-mail: joshiprachi068@gmail.com*

[2]*Assistant Professor, Department of Computer Applications, LSM Government PG College, Pithoragarh, Uttarakhand, India, E-mail: bp.pande21@gmail.com*

ABSTRACT

The world has been witnessing a grave epidemic in the current year. In December 2019, scientists realized the genesis of a novel virus *SARS-CoV-2* in Wuhan, China. Since then, this highly contagious novel virus has been causing a deadly disease in humans called coronavirus disease 2019 (COVID-19), and it was declared as a pandemic by the World Health Organization (WHO) on 11th March 2020. The fatality of the COVID-19 pandemic forced the governments all over the globe to impose and implement worldwide curfews, mass quarantines, and shutdowns known as lockdown to slow down the spread of this virus. Lockdown and associated restrictions are the very old tools for controlling epidemics. China was the first country to impose lockdown during the current pandemic in January 2020. Since then, around 90 countries implemented partial or complete lockdown and this caused approximately half of the human population to endure confinement and restrictions.

Lockdown of the substantial period affected all the realms of human life and society. There is a massive jolt of lockdown and its consequences over the economy, education, employment, transportation, manufacturing, supply chains, tourism, social, and religious rituals, gatherings, and festivals, etc. The preliminary study has revealed the fact that the three classes of our social system, say higher, middle, and lower class have been affected to different degrees. These affected groups can further be classified based on occupation, age, and environment, etc. The present chapter investigates the effects and impact of lockdown on the above-mentioned groups. Moreover, the Internet has been playing the most vital role in the current epidemic. The online activities related to the broadcasting of information, news, and updates about the pandemic, official work, education, banking, e-shopping, awareness campaigns, and entertainment, etc., are completely dependent on the internet service providers (ISPs). In India, the majority of the population is dependent on cellular networks to fulfill their Internet needs. The obvious consequence of lockdown is thus the high data consumption and load over mobile networks. The present chapter studies all the dimensions of the possible impact of lockdown on mobile networks. The current work also presents the impact and consequences of lockdown in the *Uttarakhand* province of India.

4.1 INTRODUCTION

According to the World Health Organization (WHO), a pandemic can be defined as an epidemic that occurs over a large geographical area, crossing international boundaries, and affects a substantially large mass of people [1]. Kelly [1] says that a real influenza epidemic occurs when contemporaneous transmissions take place globally. The potential impact of influenza pandemics can be visualized with two parameters: first is transmissibility, which is defined as the average number of humans infected by a single infectious carrier, and second, disease severity, which is estimated by the ratio of fatality [1]. In the present year, the whole world has been witnessing and enduring the threat of the coronavirus disease 2019 (COVID-19) pandemic. This infectious disease is caused by a novel virus known as severe acute respiratory syndrome coronavirus 2 (SARS-CoV-2). The first occurrence of this disease was identified in December 2019 in Wuhan, Hubei, China [2], and on 11th

March 2020, the WHO declared the COVID-19 outbreak a pandemic [3]. The dreadfulness of the COVID-19 enforced the governments all over the globe to adopt nationwide lockdowns to prevent the further spread of the deadly CoVs.

The term *lockdown* can be defined as a set of constraints for humans limiting their physical movements and forcing them to remain at the place they are. Such restrictions are generally imposed because of some specific risks to others or themselves had they allowed to move freely. The term lockdown is generally governed by a *prison protocol* that can only be triggered by top-level authorities and lockdown measures are initiated as preventive measures at the time of emergencies and pandemics [4]. The existence of lockdowns have been reported in the history of mankind in various forms and flavors, like fighting against pandemics; terrorism; physical, political, and technological threats; and emergencies, etc. Tools like confinement, isolation, quarantine, and social distancing practices and total lockdown have been helping to save lives and mitigating potential risks. Jahanbegloo [5] reported that the practices of isolation, confinement, and quarantine are very old and proved very helpful at the time of outbreaks of serious influenza, plague, and epidemics. Jahanbegloo [5] also shed light on lockdowns that existed in history due to nuclear accidents and terrorist attacks. Hendry [6] reported that the lockdown techniques came into existence in the late 1970s in Southern California. Williams [7] mentioned that the practice of quarantine and isolation has long been used to prevent the spread of diseases like plague, black death, sexually transmitted diseases (STDs) during World War One and the SARS epidemic.

Some scholars classify lockdown as either *total lockdown* or *partial lockdown*. In a total or complete lockdown, all activities other than the delivery of essential goods and services are suspended till further decree and the public is ordered to stay in place. A total lockdown may be adopted with a view to some potential critical situation or emergency that can cause an expeditious threat to the health or safety of the people. On the other hand, in partial lockdown, restrictions are on specific places and activities, and daily jobs can continue with precautions and care. In a partial lockdown, restrictions similar to complete lockdown can be exercised at a fixed or variable period like on every weekend, etc. In the case of COVID-19 pandemic, governments all over the globe realized the graveness of the circumstances and imposed nationwide mass quarantines, stay-at-home restrictions, curfews, and shutdowns. The idea behind such strict lockdown measures is to *'flatten the curve'* of the growth of infection advancement.

To combat the outbreak of the COVID-19 pandemic, China was the first country to impose nationwide lockdown in late January 2020 [8]. In the second week of March 2020, Italy imposed national-level mass quarantine restrictions and adopted additional lockdown provisions later. Denmark also announced a lockdown on the same week. Fiji announced the closure of schools and non-essential services in the third week of March. In the same week, France started restrictions on daily activities and Malaysia implemented partial lockdown. France announced full lockdown in the mid of March. On 25th March 2020, India announced a complete lockdown and allowed activities related to essential goods and services only. Ireland first declared shutting down of childcare activities, schools, and colleges in the second week of March and then declared nationwide total lockdown in the last week of March. By the end of March 2020, a substantial number of countries like Namibia, New Zealand, Philippines, Norway, Kuwait, Poland, Czech Republic, etc., had implemented partial or total lockdown orders. In the United Kingdom, stay-at-home orders were released in the third week of March and then re-exercised in the second week of May. In the United States (US), different orders of partial or strict lockdowns were released and updated since mid of March 2020. On the other hand, there are countries like Japan and South Korea which did not implement lockdown provisions in the current pandemic of COVID-19 [9].

The consequences of lockdown appear to be very serious. The world-wide COVID-19 lockdown is the history's one of the biggest lockdowns and it affected all the facets of human life. It directly smashed almost all of the important sectors of modern society like the economy, education, manufacturing, employment, income resources, agriculture, transporta-tion, supply chain, social, and religious ceremonies, festivals, etc. There are some indirect effects of lockdown that can be observed and contem-plated. The health of people with existing ailments, other than COVID-19 has been compromised due to the limited healthcare facilities and unavail-ability of conveyance during the lockdown. People who needed emergency care could not get the best of it. The isolation and quarantine restrictions increased fear and anxiety in people of all age groups. Social discon-nectedness catalyzed mental, emotional, and psychological health issues. People faced grave difficulties in carrying out religious and social rituals. However, there were some positive effects of lockdown like people spent time with their loved ones, emotional bonding among family members grew stronger, environmental parameters improved and natural resources rejuvenated, people spent time to develop their skills, etc.

The rest of the present chapter is organized as follows: Section 4.2 presents a brief literature survey; Section 4.3 discusses the social impact of lockdown; Section 4.4, discusses the effects of lockdown on mobile networks; in Section 4.5, we present a case study of *Uttarakhand* and the final section presents the conclusion of our research.

4.2 A BRIEF LITERATURE SURVEY

Kanitkar [10] discussed the impact on COVID-19 lockdown on the Indian economy and power sector. Kanitkar [10] estimated the economic losses in India and reported that GDP would fall by 10–31%. He [10] also discussed the power sector's demand and supply of electricity and emissions of carbon-di-oxide and found that there was a reduction of 26% in supply during the lockdown. Chaudhary et al. [11] studied the effects of lockdown on aviation, tourism, mobility, markets, etc., and their consequences on the Indian economy. Chaudhary, Sodani, and Das [11] reported heavy losses on various economic fronts and decline of GDP, however, they feel the post-pandemic scope of Indian entrepreneurship in global supply chain markets. Kumar and Dwivedi [12] studied the impact of lockdown on habits and daily routines of people. They found significant changes in Internet usage, sleep duration, eating habits, and way of living. Dutta et al. [13] discussed the impact of lockdown on the global employment sector. Kazmi et al. [14] studied the effects of lockdown on the mental health of people. Kazmi, Hasan, Talib, and Saxena [14] conducted experiments to assess various parameters of mental health like stress, anxiety, and depression and reported significant differences across gender, age, and employment status. Rutz et al. [15] studied the effects of reduced human mobility on wildlife during the worldwide lockdown. Rutz et al. [15] reported that many species of animals found peace, and quiet environment while some of them came under pressure. Kumari and Toshniwal [16] presented the impact of lockdown on air quality. Kumari and Toshniwal [16] studied the concentration of a few air pollutants in three Indian cities, namely, Delhi, Mumbai, and Singrauli for two different phases: pre-lockdown and post-lockdown. They observed significant improvement in the air quality of Delhi and Mumbai during the lockdown period, although there was not much improvement in air pollution in Singrauli because of the active coal-based power plants [16]. Lokhandwala and Gautam [17] presented the effects of COVID-19 lockdown on the environment and observed the improvement in air and water quality. Lokhandwala and Gautam [17] reported that the air quality index declined to two figures in all the states of India and rivers became clearer, which made clear visibility of marine life. Singh et al. [18] studied the consequences of COVID-19 lockdown on the mental health of children and young adults. Singh et al. [18] felt the need to develop sound strategies to assist social, mental, and psychological health issues of children and teenagers during the epidemic and post epidemic.

4.3   SOCIAL IMPACT OF LOCKDOWN

In the present section, various impacts of lockdown have been discussed. The present year has witnessed history's most grand lockdown ever, and its consequences are indeed grave. Almost every realm of human's social life went through a substantial jolt of hibernation and inconsistency. The impact of COVID-19 lockdown had different effects on the three social classes, say upper, middle, and lower. Undoubtedly, this massive lockdown has affected our social, economic, mental, psychological lives greatly, but some positive effects of it have also been realized.

4.3.1   AREAS AFFECTED BY LOCKDOWN

The COVID-19, with the status of a grave pandemic, induced a deep fear, confusion, and dreadful impact on us that we have never felt before. It has left its marks in a multitude of realms. Some of the major areas which suffered a deep impact of the COVID-19 lockdown are economy, educational system, agriculture, hospitality, and tourism, health, manufacturing, transportation, supply chain, etc.

4.3.1.1   ECONOMY

Due to the worldwide lockdown in 2020, the global economy has been suffering from a very tough phase and how much recession it may have to face in the near future cannot be estimated now. According to the United Nations Conference on Trade and Development (UNCTAD) report, the global economy will fall by 4.3% in the year 2020 [19]. Since many countries have been facing economic fall this year and they will hardly be able to elevate it in the next year, the whole world is expected to lags behind by almost 21 years. The International Air Transport Association (IATA) has predicted that airline may suffer a loss of $113 billion in the current year [20]. Airlines make a huge contribution to the world's economy; it can easily be inferred that the world economy has been deeply shocked by this epidemic. The COVID-19 lockdown impacted the Indian economy adversely, and it has been speculated by the *UNCTAD* report that the Indian economy will be declined by 5.9% in the present year [19].

4.3.1.2   *EDUCATION*

The education system is the foundation of every country's future. This lockdown deeply affected the global educational system as schools, colleges, and universities across the globe have been kept closed until further instructions by the administrative authorities. The United Nations Educational, Scientific, and Cultural Organization (UNESCO) prepared a report on pre-primary, primary, lower secondary, and upper secondary students all over the world and presented the ratio of students directly affected by COVID-19 lockdown. According to *UNESCO*, in the month of April 2020, 90% of total enrolled students were directly affected by the lockdown, and in the mid-June, this ratio was decreased to 56.7%. According to the current data of the last week of September 2020, the ratio has fallen to 48.6%, because some countries resumed schools and colleges in unaffected areas [21]. This epidemic suddenly worsened the schedule of every educational institute. The schools, colleges, and universities started classes in online mode, and they had to cancel or postpone all examinations. As a result, the traditional classroom teaching and learning practices were discontinued and all the educational institutes turned towards online teaching and learning paradigm to sustain the educational system. Ultimately, such consequences of the COVID-19 lockdown brought a great challenge for the teachers, students as well as parents. The compulsion of teaching in online mode put a lot of pressure on teachers, especially on those who were not comfortable with information technology (IT) tools and practices. Since the early days of lockdown, digital platforms like *Zoom, Google Meet, Google Classroom, Microsoft Teams, WhatsApp groups, YouTube Live*, etc., have been in extensive demand. One has to be technologically proficient to deal with these digital tools and technologies and to prepare the e-contents like videos, slides, PDFs, etc., to support their lectures. In India, several initiatives have been taken to educate children from secondary to higher education. The Ministry of Human Resource Development (MHRD) released online portals, TV channels, and radio channels to provide free access to online study material. Therefore, it can be said that this lockdown brought negative impacts on the education system, but there were some positive effects as well in the sense of transparency and enhancement of technical skills [22, 23]. The consequences of this lockdown are going to be worst for the students whose parents would not be able to pay their school fees because of the unemployment due to the economic crisis.

4.3.1.3 AGRICULTURE

Those people are the most affected by this lockdown who make a huge contribution to the Indian economy informally, our farmers. India's almost 59% of citizens directly depend on agriculture for their survival. India produces many crops which are classified into two major categories according to the climate conditions: *Rabi* and *Kharif*. The *Rabi* crops are sown at the start of winters (October-November) and the *Kharif* crops get sown at the start of the monsoon (May-June). The *Rabi* crops are harvested in the month of April, and it includes wheat, pulses, mustard barley, oats, etc. The decision of complete lockdown in India from the third week of March 2020 directly affected the harvesting of the *Rabi* crops. Optimum harvesting requires huge manpower and machinery, which would not be possible due to the strict lockdown restrictions. Therefore, the farmers themselves harvested the entire crop and carried it to the *Mandi* through their own channels as the public transportation was closed. They did not get the profit as much as they would have gotten in normal days [24]. The situation was worse for those who grew perishable items, as they could not get any buyer [25]. The COVID-19 lockdown severely impacted the agriculture, farmers suffered heavy losses, the economy of the country fell, food insecurities were greatly increased, and the employment of many daily workers has been lost, which caused many of them to starve [26, 27].

4.3.1.4   HOSPITALITY AND TOURISM

This COVID-19 lockdown caused a profound impact on the hospitality and tourism sector globally. The hospitality industries have been going through a very bad time. Small businesses like restaurants, pubs, bars, etc., all suffered heavy losses as the whole hospitality sector remained closed during the lockdown, and even months after lockdown, people have been avoiding to visit such places [28]. Moreover, almost all countries banned international flights and public transportation to prevent the spread of the virus. People were allowed to travel only for some critical reason. In India, there are many parts which have been operating as a tourism hub, and the lockdown was proved as a major setback to the tourism industry. Usually, hotels and resorts at such tourist hubs used to be booked a few months in advance. The sudden announcement of lockdown made the customers cancel their advanced bookings. This caused the tourism and hospitality industries to bear very severe damage [29]. According to the National Restaurant Association of India (NRAI), the hospitality sector provides employment to around 7 lac people in India [30], and if the hospitality industry cannot operate normally, such a large number of people may go under the threat of unemployment. Unfortunately, as a result of this lockdown, many hotel staff were fired from their work, or their salaries were withheld for a few months. Overall, the employees who have been working in this sector got affected to different degrees: some of them lost their jobs permanently, some of them lost their job for a temporary period, and some had to bear with the deducted salary [31]. To overcome such unpleasant consequences of the COVID-19 lockdown, both business personnel and consumers should be mutually sensitive, respectful, and should help each other to grow at the time of this crisis.

4.3.1.5   HEALTH

The COVID-19 epidemic and worldwide lockdown induced fear and anxiety in people of almost all age groups. The circumstances brought moderate to substantial mental, emotional, and psychological consequences. Singh et al. reported that the lockdown led to an adverse impact on the mental and psychological health of children and adolescents [18]. The authors realized the pivotal need for implementing well-thought plans to address the psychosocial and mental health needs of people in the tenure of COVID-19 epidemic and up to a substantial period after it. The physical and mental health of people that belong to the poor social-economic groups like laborers, migrant workers, homeless refugees, asylum seekers, etc., was compromised more severely during this lockdown. Moreover, the general public endured grave difficulties in accessing normal health services and treatment assistance during the lockdown. There has been no or very limited arrangements to assist and cure people with pre-existing ailments, health issues other than COVID-19, emergencies like accidents, delivery cases, etc.

4.3.1.6   MANUFACTURING, TRANSPORTATION, AND SUPPLY CHAIN

The manufacturing companies contribute almost 20% of GDP, and the strict lockdown restrictions caused manufacturing companies to endure massive losses. The manufacturing processes were not exempted from the lockdown during the initial phases. Also, it was greatly affected by the lack of available transportation at the time of lockdown. Demand was declined due to the closure of allied businesses. The pharmaceutical companies are a very good example of the adverse effects of lockdown. In the initial phase of lockdown, public transportation was almost completely banned. Transportation-related to only essential goods and services was allowed, and agents needed passes for movement. The public transportation industry was hit badly by huge losses, and consequently, the nation's economy was compromised. It has been the very first time in history that the whole rail network was shut down in India. Every means of transportation suffered a sharp decline in income, from rickshaw to airline. In the unlock process after lockdown, government-approved permission to the transport industry to run with the capacity of 50% passengers only. Obviously, the passengers had to pay the double fare. The supply chain was also affected due to the lesser demand, unavailability of transportation, nil manufacturing, shut down of distributers and shopkeepers. Eventually, the black marketing of required items was started and they were sold at an arbitrary rate.

4.3.2 IMPACT OF LOCKDOWN ON THE THREE SOCIAL CLASSES

Our society has not been impacted equally by the COVID-19 lockdown. The lockdown made marks of different degrees to the three different classes, say lower, middle, and higher class.

4.3.2.1 LOWER CLASS FAMILIES DURING LOCKDOWN

India's COVID-19 lockdown hit daily wages and migrant workers very hard. Casual laborers make up about 90% of India's workforce. The declaration of total lockdown sat all factories and businesses non-functional. Consequently, the lower-class people were suddenly thrown in front of the burning question of bread and butter. They did possess limited stockpile which was totally insufficient to meet their future food requirements. This critical situation forced the migrant laborers to return to their hometowns. The unavailability of public transportation raised their problems to even more extent. Consequently, many migrant laborers decided to cover huge distances to their hometowns on foot. Such unfortunate compulsion is almost impossible for other classes even to imagine. Lower-class people put their lives on stake to combat the fear of starvation. Eventually, the state and central government announced some relief packages and some NGOs also came forward to arrange and deliver cooked food kits to such needy people during the lockdown.

4.3.2.2 MIDDLE-CLASS FAMILIES DURING LOCKDOWN

The middle-class family is a class which contributes a substantial share to the economy of the whole nation. Different people hold different opinions about the income of middle-class people. The middle class is further divided into two sub-classes: the upper-middle class and the lower-middle class. The lower-middle-class people get moderate salary and many of them either lost their jobs temporally or suffered heavy deductions in their income. They possessed limited stockpile that could serve them for a few months only. They could not avail benefits provided by the government at the time of lockdown as they fall outside the below poverty level (BPL) criterion. Their self-esteem might refuse to get benefited by the free food kits provided by government agencies or NGOs. Therefore, they had to live by mitigating their needs and endeavoring to find some else income sources [32]. On the other hand, the upper-middle-class people draw a relatively good salary and their socio-economic status is stable. Some of them might lose their jobs temporarily or permanently, and some suffered moderate to heavy deductions in income too. But their livelihood was not suffered much during the lockdown, and their lives are being resumed to normal now.

4.3.2.3   HIGH-CLASS FAMILIES DURING LOCKDOWN

High-class families include self-employed people possessing large businesses, celebrities, politicians, real estate licensees, etc. There were no financial and logistic hurdles for this class, but they were affected mentally, physically, and socially during the COVID-19 lockdown. They had been stuck to the rich social environment and being in confinement all of the time brought emotional, mental, and psychological health implications in their ways. It is said that many of them were addicted to gambling and sedatives.

On the other hand, the COVID-19 lockdown bestowed some positive opportunities to middle-class and high-class people: they spent quality time with their family and loved ones, they endeavored to develop their skills, they concentrated on their health and hobbies. The overall impact of lockdown on the three social classes can be categorized by Table 4.1.

TABLE 4.1   Social Impact of Lockdown

The restrictions have been easy to tolerate for:	• Younger, mentally, and physically resilient elements of society.
	• White-collar workers who can work remotely.
	• Relatively affluent with access to food, space, and gardens.
	• Those living with others they love and that they are loved by.
	• Those that can share the load of parenting and care for others.
The restrictions have been harder to tolerate for:	• Those with pre-existing mental and physical health issues.
	• Health and social care workers
	• Those that lost employment.
	• City-based renters with less access to food, space, and gardens
	• Those who solely responsible for caring for their children and parents.

4.3.3 IMPACT OF LOCKDOWN ON ENVIRONMENT

Undoubtedly, the COVID-19 lockdown descended as a boon for the environment. Restrictions on humans mitigated their interference on nature, and this gave an excellent chance for its revival and rejuvenation.

4.3.3.1 AIR QUALITY

During the COVID-19 lockdown, the government instructed strict guidelines to follow social distancing and stay-at-home restrictions. Public transportation was completely banned and this made a substantial difference in air quality. It has been ascertained that lockdown affected outdoor air pollution and it was declined up to a remarkable level. In India about, 90% of the metropolitan cities had very poor air quality before lockdown, causing many diseases like lung cancer, asthma, heart disease, premature deaths, etc. The particulates, ozone, nitrogen dioxide, Sulfur-dioxide are the major components that cause air pollution. It has been observed that during and just after the lockdown, very lesser amounts of these harmful gases were measured because of the ceased road traffic and closure of smoke emanating factories. According to a study, the lockdown improved air quality index very well in the early days, say from 100 to 61. But as soon as some relaxation was given to the public, it again reached to 71 [33].

4.3.3.2 WATER QUALITY

The factories became non-operational during the lockdown, and consequently, the practice of throwing industrial waste into rivers was ceased. The daily, ritual, and religious activities of people on river banks were also completely stopped. Such circumstances definitely caused a positive impact on their water quality. During the COVID-19 lockdown, the water quality of rivers and other reservoirs has improved drastically. The water quality of the *Ganges* has improved by around 40–50% during the lockdown, and the locals reported that the water is much cleaner than before.

4.3.3.3 OZONE LAYER

The Ozone layer envelops the earth and protects living things on the planet's surface by the ultraviolet (UV) radiation. Depletion of the Ozone layer is due to many factors, the most dominant of them is the release of *Chlorine* from *Chlorofluorocarbons* which destroys the Ozone. The scientists confirmed that the largest hole in the ozone layer was healed by itself amid the worldwide COVID-19 lockdown.

4.4 IMPACT OF LOCKDOWN ON MOBILE NETWORKS

The stay-at-home orders during the outbreak of the COVID-19 pandemic gave birth to the need of adopting *work from home* culture and practices in most of the job profiles. The Internet played a vital role to assist such crisis. People spent their leisure time in entertainment, surfing, learning, online gaming, social media, etc., and consequently, all of them needed access to the Internet at a reasonable speed. In India, access to the Internet through cellular networks hold the largest share of the whole internet service provider (ISP) industry. The COVID-19 lockdown was a time in the whole world when most of the people were completely dependent on their mobile networks to continue their professional and personal affairs. The Ministry of Home Affairs (MHA), Government of India exempted public and private sector companies dealing in telecommunications, Internet services, IT-enabled services to remain closed during the lockdown [34]. During the lockdown, the ISPs noticed an increase of 10% in the overall data usage, with 20% spike in the views of streaming platforms [35]. The cellular operators association of India (COAI) wrote a letter to all the telecom companies to reduce the data quality for streaming platforms like Netflix, Hotstar, and Zee5, etc., so that consumers may not feel any interruption despite the increased traffic. Many top telecom companies switched their data from high definition (HD) to standard definition (SD) mode, due to the increased strain on mobile networks. However, some states did not approve working permission of field employees of telecom companies. Such circumstances brought difficulties for them to continue services smoothly. Eventually, the Department of Telecommunication (DoT) requested every state to cooperate so that people can get uninterrupted Internet facilities. The telecom companies also recorded a lesser number of new subscriptions, and many telecom companies changed their tariff plans to provide better services at a low cost.

4.5 IMPACT OF COVID-19 LOCKDOWN ON THE LIVES OF *UTTARAKHAND'S* RESIDENTS: A CASE STUDY

The present case study has been undertaken to assess the impact of lockdown on the lives of people of the *Uttarakhand* state of India at a surface level. A survey questionnaire was designed asking people about their opinions and experiences on social, financial, and psychological fronts.

4.5.1 MATERIAL AND METHODOLOGY

Because of the mandate of maintaining social distancing in the current pandemic, the survey could be conducted only in the online mode. Therefore, an online survey was designed and disseminated with the help of IT tools. A semi-structured questionnaire was prepared in *Google Forms* and was circulated among the respondents, the residents on *Uttarakhand* only via social media platforms like Facebook and LinkedIn posts, WhatsApp chats, e-mails, etc. The survey was conducted for a period of two weeks, say from 15th September to 30th September 2020 and around 500 people from the *Uttarakhand* state participated in this online survey. This survey was open to all age groups.

4.5.2 QUESTIONS AND RESPONSES

Among the participants who took the survey, the majority of the respondents belongs to middle-class family (89.7%), some of them belong to low-class family (7.9%) and few of them fall in high-class family (2.4%). The questions and the responses for them are depicted in Figure 4.1.

4.6 CONCLUSIONS

The whole world witnessed an unprecedented lockdown in 2020. Declaration and implementation of strict restrictions regarding mass quarantine and confinement, shut down of all non-essential activities, curfews, etc., by governments all over the globe make the lockdown caused by COVID-19 as the biggest lockdown in the history of human civilization. Almost half of the human population on the earth endured confinement and allied restrictions as more than 90 countries declared lockdown. China initiated lockdown provisions in late January 2020, and by the end of March, almost all affected countries declared lockdown restrictions of varying degrees. The consequences of such a massive lockdown have put deep marks on almost all the spheres of human life. This lockdown directly affected all the vital elements of modern society, and the pillars of the economy, education, health, social stability, and progress went through a substantial jolt of adversity and inconsistency. Some indirect effects of lockdown have also been realized like limited availability and assistance of health care personnel; mental, emotional, and psychological health implications; difficulties in carrying out social and religious rituals, inclination towards addictive habits, etc. However, the lockdown brought some opportunities like people got the chance to spent lives with family and loved ones, to enhance their knowledge and skills, to practice their hobbies, etc. The lockdown definitely proved a boon for the environment and nature.

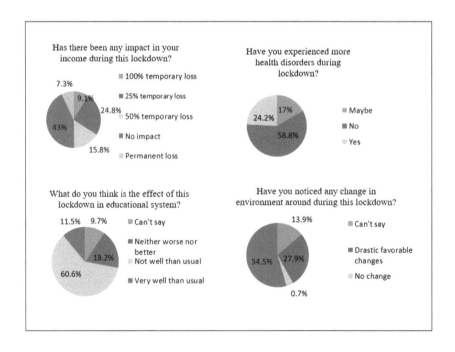

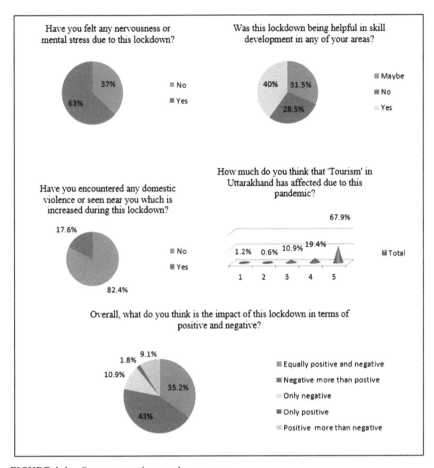

FIGURE 4.1   Survey questions and responses.

The analysis of the survey on the residents of Uttarakhand indicated that the income of most of the residents stood unaffected during the lockdown while very few borne temporary losses. Most residents had lesser health disorders, but many of them experienced anxiety and nervousness. Most people believe that there were slight favorable changes in the environment, but the tourism sector of Uttarakhand has affected greatly. Many people believe that lockdown implications on the education system cannot be considered healthy, although it provided opportunities for skill development. A great percentage of Uttarakhand residents did not notice any domestic violence during the lockdown tenure. The difference between the number of people who believe lockdown brought only negative impacts and who believe it brought both negative and positive impacts is not substantial.

KEYWORDS

- **coronavirus disease 2019**
- **lockdown**
- **quarantine**
- **severe acute respiratory syndrome coronavirus 2**
- **sexually transmitted diseases**
- **social distancing**
- **United Nation Conference on Trade and Development**
- **World Health Organization**

REFERENCES

1. Kelly, H., (2011). The classical definition of a pandemic is not elusive. *Bulletin of the World Health Organization, 89*, 540, 541.
2. *Coronavirus Disease 2019*, (2020). Wikipedia. https://en.wikipedia.org/wiki/Coronavirus_disease_2019 (accessed on 22 June 2021).
3. *Rolling Updates on Coronavirus Disease (COVID-19)*, (2020). World Health Organization (WHO). https://www.who.int/emergencies/diseases/novel-coronavirus-2019/events-as-they-happen (accessed on 22 June 2021).
4. *Lockdown*, (2020). Wikipedia. https://en.wikipedia.org/wiki/Lockdown (accessed on 22 June 2021).
5. Jahanbegloo, R., (2020). *Life Lessons from the History of Lockdowns*. Mint., https://www.livemint.com/news/india/life-lessons-from-the-history-of-lockdowns-11585312953744.html (accessed on 22 June 2021).
6. Hendry, J. A., (2020). *The Origin of Lockdown*. ALICE Training Institute. https://www.alicetraining.com/resources/item/origin-of-lockdown/ (accessed on 22 June 2021).
7. Williams, S., (2020). *Coronavirus: How Quarantine has Fought Disease Through the Ages*. BBC News. https://www.bbc.com/news/world-51308542 (accessed on 22 June 2021).
8. Levenson, M., (2020). *Scale of China's Wuhan Shutdown is Believed to be Without Precedent*. The New York Times. ISSN: 0362-4331.

9. *COVID-19 Pandemic Lockdowns,* (2020). Wikipedia. https://en.wikipedia.org/wiki/ COVID-19_pandemic_lockdowns (accessed on 22 June 2021).

10. Kanitkar, T., (2020). The COVID-19 lockdown in India: Impacts on the economy and the power sector. *Global Transitions, 2,* 150–156.

11. Chaudhary, M., Sodani, P. R., & Das, S., (2020). Effect of COVID-19 on economy in India: Some reflections for policy and program. *Journal of Health Management.* doi: 10.1177/0972063420935541.

12. Kumar, M., & Dwivedi, S., (2020). Impact of coronavirus imposed lockdown on Indian population and their habits. *International Journal of Science and Healthcare Research, 5*(2), 88–97.

13. Dutta, J., Mitra, A., Zaman, S., & Mitra, A., (2020). Lockdown and beyond: Impact of COVID-19 pandemic on global employment sector with special reference to India. *NUJS Journal of Regulatory Studies,* 1–4.

14. Kazmi, S. S. H., Hasan, K., Talib, S., & Saxena, S., (2020). *COVID-19 and Lockdown: A Study on the Impact on Mental health.* Research Gate. doi: 10.13140/ RG.2.2.21490.04808.

15. Rutz, C., Loretto, M., Bates, A. E., et al., (2020). COVID-19 lockdown allows researchers to quantify the effects of human activity on wildlife. *Nat. Ecol. Evol., 4,* 1156–1159.

16. Kumari, P., & Toshniwal, D., (2020). Impact of lockdown measures during COVID-19 on air quality: A case study of India. *International Journal of Environmental Health Research.* doi: 10.1080/09603123.2020.1778646.

17. Lokhandwala, S., & Gautam, P., (2020). Indirect impact of COVID-19 on environment: A brief study in Indian context. *Elsevier Public Health Emergency Collection, 188,* 109807, doi: 10.1016/j.envres.2020.109807.

18. Singh, S., Roy, D., Sinha, K., Parveen, S., Sharma, G., & Joshi, G., (2020). Impact of COVID-19 and lockdown on mental health of children and adolescents: A narrative review with recommendations. *Psychiatry Research, 293,* 113429. doi: 10.1016/j. psychres.2020.113429.

19. PTI, (2020). *Amid COVID-19 Impact, Indian Economy Forecast to Contract 5.9% in 2020: U.N.* The Hindu. https://www.thehindu.com/news/national/amid-covid-19-impact-indian-economy-forecast-to-contract-59-in-2020-un/article32675047.ece (accessed on 22 June 2021).

20. Appleton, P., (2020). *Potential for Revenue Losses of $113bn Due to COVID-19 "Crisis."* International Air Transport Association (IATA). https:// airlines.iata.org/news/potential-for-revenue-losses-of-113bn-due-to-covid-19-%E2%80%9Ccrisis%E2%80%9D (accessed on 22 June 2021).

21. *Education: From Disruption to Recovery,* (2020). UNESCO. https://en.unesco.org/ covid19/educationresponse (accessed on 22 June 2021).

22. Jena, P. K., (2020). Impact of pandemic COVID-19 on education in India. *International Journal of Current Research, 12*(7), 12582–12586.

23. Jadhav, V. R., Bagul, T. D., & Aswale, S. R., (2020). COVID-19 era: Students' role to look at problems in education system during lockdown issues in Maharashtra, India. *International Journal of Research and Review, 7*(5), 328–331.

24. Vasudeva, V., & Jebaraj, P., (2020). *Farming Under Lockdown: Short on Laborer's, a Long Harvest.* The Hindu. https://www.thehindu.com/news/national/other-states/

farming-under-lockdown-short-on-labourers-a-long-harvest/article31370176.ece (accessed on 22 June 2021).

25. *Supply Chain Breaks, Farmers in Distress,* (2020). The New Indian Express. https:// www.newindianexpress.com/states/karnataka/2020/apr/01/supply-chain-breaks-farmers-in-distress-2124166.html (accessed on 22 June 2021).

26. Rawal, V., Kumar, M., Verma, A., & Pais, J., (2020). *COVID-19 Lockdown: Impact on Agriculture and Rural Economy.* Society for social and economic research. ISBN: 978-81-937148-8-1.

27. Arumugam, U., Kanagavalli, G., & Manida, M., (2020). COVID-19: Impact of agriculture in India. *Aegaeum, 8*(5), 480–488.

28. Gössling, S., Scott, D., & Hall, C. M., (2020). Pandemics, tourism and global change: A rapid assessment of COVID-19. *Journal of Sustainable Tourism.* doi: 10.1080/09669582.2020.1758708.

29. Şengel, Ü, Çevrimkaya, M., Işkın, M., & Zengin, B., (2020). The effects of novel coronavirus (COVID-19) on hospitality industry: A case study. *Journal of Tourism and Gastronomy Studies, 8*(3). 1646-1667. doi: 10.21325/jotags.2020.626.

30. Priyadershini, S., (2020). *The Future of Eating out in India as Lockdown Restrictions Loosen.* The Hindu. https://www.thehindu.com/life-and-style/food/as-lockdown-restrictions-loosen-indias-restaurant-industry-discusses-ways-to-encourage-contact-less-dining-and-stay-in-business/article31557920.ece (accessed on 22 June 2021).

31. Nainar, N., (2020). *COVID-19 Takes a Toll on Hotel Industry.* The Hindu. https:// www.thehindu.com/news/cities/Tiruchirapalli/covid-10-takes-a-toll-on-hotel-industry/article31333243.ece (accessed on 22 June 2021).

32. Choubey, M. K., (2020). Corona pandemic and troubles of Indian middle class. *Jamshedpur Research Review*, 1–6. doi: 10.13140/RG.2.2.13461.14565.

33. Deepika, K. C., (2020). *Improvement in Air Quality Wears Down with Relaxation in Lockdown.* The Hindu. https://www.thehindu.com/news/cities/bangalore/improvement-in-air-quality-wears-down-with-relaxation-in-lockdown/article31742164.ece (accessed on 22 June 2021).

34. Guidelines, (2020). *Ministry of Home Affairs (MHA)*, https://www.mha.gov.in/sites/default/files/Guidelines.pdf (accessed on 22 June 2021).

35. Mankotia, A. S., (2020). *Mobile Internet Usage Increases Just 10% Since Lockdown.* The Economic Times. https://economictimes.indiatimes.com/tech/internet/mobile-internet-usage-increases-just-10-since-lockdown/articleshow/74920799. cms?from=mdr (accessed on 22 June 2021).

CHAPTER 5

Forecasting the Damage Caused by COVID-19 Using Time Series Analysis and Study of the Consequence of Preventive Measures for Spread Control

BASUDEBA BEHERA, UJJWAL GUPTA, and SAGAR RAI

Department of Electronics and Communication Engineering, National Institute of Technology Jamshedpur, Jharkhand, India, Tel.: +91-8812016250, E-mail: basudeb.ece@nitjsr.ac.in (B. Behera)

ABSTRACT

The World Health Organization (WHO) has declared the outbreak of the novel coronavirus (nCoV) as a pandemic. With the exponentially increasing cases of the virus throughout the world, forecasting the number can be an important aspect for further data analysis of the growth of the confirmed cases, deaths, and recoveries around the globe. To accomplish this task of forecasting the growing/depreciating count of confirmed cases, deaths, and recoveries, we have used the open-source tool of Prophet made available by Facebook. In the absence of any vaccine currently present, different countries have resorted to various preventive measures for slowing the rate of spread of the virus and flattening the curve of confirmed cases. A comparative study between two nations, China and Italy, has been performed for seeing the effect of preventive measures on the growth of the virus. This chapter can serve as a model for organizations to understand the exponential growth and need for implementation of preventive measures for subsidizing the spread.

5.1   INTRODUCTION

The novel coronavirus (nCoV) or COVID-19 has originated from the Hubei province in China and has spread rapidly around the globe. The World Health Organization (WHO) has declared the disease as a pandemic. As there is no vaccine present for the disease, various countries have resorted to different means for control of the spread. As of 2nd April 2020, there are around 1,015,065 confirmed cases of the nCoV or COVID-19 in the world. Coronaviruses are enveloped positive-sense, non-segmented RNA viruses belonging to the Coronaviridae family and the Nidovirales order and widely distributed in humans and other mammals [2]. The virus causes a range of symptoms ranging from fever, difficulty in breathing, dry cough, fatigue, and bilateral lung infiltration in severe cases. Some of the patients showed non-respiratory symptoms including nausea, diarrhea, and vomiting. Most of the patients are directly or indirectly connected to the people visiting the live animal market in Huanan. The Chinese health authority initially suggested that the patients had tested negative for the earlier known viruses and bacteria but tested positive for the nCoV [3]. In contrast to the earlier studies, the nCoV spreads from human to human as found in Ref. [3]. During the pandemic period in which the cases of the disease are rising rapidly each day, forecasting is of utmost importance for tackling the disease with limited resources [1].

In the first part of our work, time series analysis is used to forecast the number of cases around the world based on the available data. This work also forecast the number of new cases per day in the world. The FBProphet Time-Series analysis model is used in the forecasting part. Then this work forecasts the number of cases in Italy and China based on the data available for the respective countries. The data used for the analysis has been taken from the 2019 novel coronavirus COVID-19 (2019nCoV) Data Repository by Johns Hopkins CSSE on GitHub [4]. In the second part, we have studied and compared the present death rates in the World, Italy, and China to see how the different kinds of preventive measures implemented by different countries can help slow the spread of the infection. The death rates have been calculated based on only use the cases that are concluded that are either dead or recovered since we do not know the outcome of the cases that are still in process. We calculate the death rate as (death)/(death + recovered). The chapter serves as a predicted forecast of total cases throughout the world with a comparison of confirmed cases and deaths in Italy and China to see how the various preventive measures affect the spread of the infection.

5.2   LITERARY SURVEY

The first application of a time-series analysis model dates back to the 1920s and 1930s to the work of Yule and Walker [5]. Auto-Regressive models were the building blocks for the time-series analysis models. The Box-Jenkins Approach [6] towards statistics brought revolutionary changes to the concept of time-series analysis. The most commonly used autoregressive models being the ARIMA (autoregressive integrated moving average) [7]. But it requires fine-tuning with forecasting over the period of data. Recently Facebook introduced a new tool for forecasting name Prophet [8], which is comparatively easy to use and predicts more relevant results. This tool was originally introduced for high-quality business forecasts [8].

The center of the study of this chapter revolves around the forecasting of the increasing or decreasing trend of the new cases of COVID-19. The impact and origin trace back to China, where the spread was exponential [9], but had recently decreased [4] due to the preventive measures been taken by the country at the right time. The Prophet model is used to forecast the number of cases in the near future which can help us to implement preventive measures to reduce the spread of this pandemic.

5.3   PROPOSED METHOD

The most important part of training a model is the data we feed in. For this work, we have used the data repository of the 2019 nCoV dashboard operated by the Johns Hopkins University Center for Systems Science and Engineering (JHU CSSE) [4]. It had used WHO, Hong Kong Department of Health, National Health Commission of the People's Republic of China (NHC), WorldoMeters, and other authenticated source for creating this data repository. It has eight columns as shown in Figure 5.1, with a well description province, country, latitude, longitude, date, confirmed positive cases, confirmed deaths, and confirmed recovered cases.

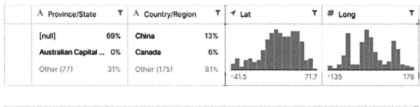

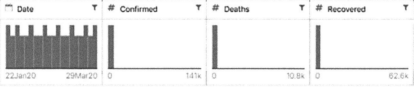

FIGURE 5.1   Statistical view of the data used.

Pre-processing of the key requirement for fitting the data into a model. The fields which would serve as the basis of input for our model would be the date, confirmed positive cases, confirmed deaths, and confirmed recovered cases. We need to filter out the other data. The other parameters namely, province, state, latitude, and longitude helped us visualize the spread of COVID-19 (Figure 5.2). Features of province and state helps in the forecasting of a particular country which is been explained later in the chapter. The cleaned data served as the input for our time series model.

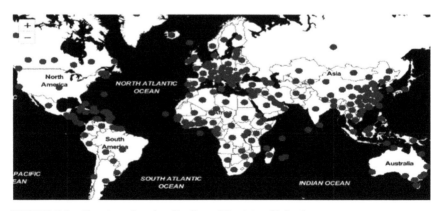

FIGURE 5.2   Country-wise visualization of the data of infections.

5.3.1   TIME SERIES ANALYSIS

Time series analysis is the study of the data and extracting information from that data [10]. Its main objective is to carefully study the past observations of a time series and create an appropriate model which describes the inherent structure of that series. Thus, the study of time series analysis can be described as an act of predicting the future by understanding the past [11]. It serves as a very important mode of study in fields of predicting future trends. The most common area of usage being the economic and sales forecasting, stock market analysis, budgetary analysis, census analysis, etc., and producing satisfactory results.

The time-series analysis can be widely categorized into two domains [12]:

- time-domain; and
- frequency-domain.

Usage of Fourier transform to convert the time series into a weighted sum of sinusoids forms the building blocks of frequency-domain methods [13]. The periodic components of the time series are calculated through the spectral analysis. The more commonly used form of time-series analysis apply to the Time-domain analysis method. In our research, we have also used the Time-domain method.

5.3.2   FB PROPHET

Prophet tool is used for the implementation of time series analysis using time-domain methods. This is an open-source tool made available by Facebook for predicting the future result. It can be used both in Python and R. It is forecasting model designed mainly for handling common features of business time series. Unlike the then used ARIMA model, it is less complex and has more features for better computations. It is built with intuitive parameters that can be changed without understanding the underlying model information, making it easier to use [14].

A time-series model can be decomposed into three main components:

- trend;
- seasonality;
- holidays.

These form the building blocks of the Prophet tool. It can be represented using the following Eqn. (1) [14]:

$$y(t) = g(t) + s(t) + h(t) + \varepsilon t \qquad (1)$$

where; g(t) represents the trend function; s(t) represents the periodic changes or the seasonality; h(t) represents the effects of holidays; and εt is nothing but the error term.

5.3.2.1   TREND FUNCTION

It is basically the fitting of the model over a trend or a non-periodic part of time-series. It has two important components:

1. **Saturation Growth:** It is nothing but the study of the growth and its extremities. It gives the upper bounds for a non-linear growing model. There can be two types of growing models: linear and logistic [14].
2. **Change Points:** Growth models can encounter sudden changes due to an unforeseen event. At such cases, growth rates need to be changed.

5.3.2.2   SEASONALITY FUNCTION

Seasonality's are the periodic changes which are observed over the course of time. Generally, a dataset represents repeated changes in the course of time. For fitting such kinds of effects, specific seasonality functions need to be added to the model. This can be approximated using the following equations [14]:

$$s(t) = \sum_{n=1}^{N} \left(a_n \cos\left(\frac{2\pi nt}{P} \right) + b_n \sin\left(\frac{2\pi nt}{P} \right) \right) \qquad (2)$$

where; P is the period where parameters $[a_1, b_1, \ldots, a_N, b_N]$ are needed to be estimated for a given N to model seasonality. Tuning of the parameter N is used for modeling the changes.

5.3.2.3   HOLIDAYS

These are the predicted shocks to the time series model. They are well known from before and setting the changes for that day helps in tabulating the data more easily. Prophet allows the user to manually enter a custom list of holidays and events for fitting it into the model.

Taking this model as the basis of our study, we have tried to predict the increase/decrease in the number of cases of COVID-19 all over the world. A Growth saturating Prophet model is being used here. The source of input is the dataset filtered by us. The key columns are the date, confirmed positive cases, confirmed deaths, and confirmed recovered cases. The model studies the data for the last 68 days and helps to forecast results for the upcoming 14 days. Python libraries like matplotlib, seaborn, and plotly are been used for the proper visualization of the data.

The predicted data is being compared with the originally growing cases around the globe, and the results are commendable (Table 5.1).

TABLE 5.1   Comparison of Forecasted Values vs. Actual Values [17]

SL. No.	Date	Actual Confirmed Cases	Predicted Confirmed Cases	Error Percentage (%)
1.	30-03-2020	784,741	795,670	1.39%
2.	31-03-2020	858,361	869,758	1.33%
3.	01-04-2020	935,232	945,814	1.13%
4.	02-04-2020	1,015,096	1,025,247	1.00%
5.	03-04-2020	1,116,662	1,110,453	0.56%

5.4   EXPERIMENT AND RESULTS

5.4.1   EXPERIMENTAL SETTINGS

The size of the input data available in the data source [4] consisted of 17,136 rows and eight columns namely Date, Province, Country, Latitude, Longitude, Confirmed, Death, and Recovered. The province data had a high number of missing values and hence was dropped for analysis. We grouped the data by date and did a sum of the confirmed, death, and recovered values for our general analysis and forecasting. We also created different data frames country-wise for two countries, namely Italy and China, for the comparison of the forecast and death rates in the two countries and the death rate, in general, to see how different preventive measures affect the growth and spread of the infection in the surroundings.

5.4.2   EXPERIMENTAL RESULTS

Firstly, we used Seaborn and Plotly libraries for plotting the total number of confirmed cases around the world. We saw that the total cases were increasing exponentially worldwide, as shown in Figure 5.3.

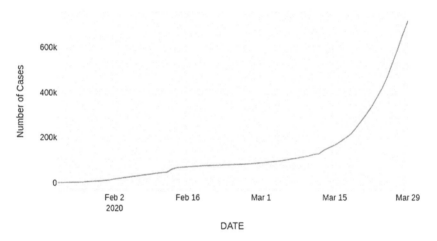

FIGURE 5.3   Total number of confirmed cases day-wise.

We then calculate the infection death rate, by using the cases which have reached their conclusion that is the patients have either recovered or died as we are not sure about the conclusion of the remaining cases. We calculate the death rate as given in Eqn. (3).

$$death\_rate = (number\ of\ deaths)/(number\ of\ deaths + number\ of\ patients\ recovered) \tag{3}$$

The death rate is then plotted on a graph as shown in Figure 5.4. We observe that the death rate initially increased to about 55% and then decreased exponentially and is currently on an increase. We find that the present death rate is around 19%. The death rate here is indicative of how many concluded cases (recovered cases or deaths) have resulted in a fatality. The calculation of the death rate does not include the cases which are still under treatment.

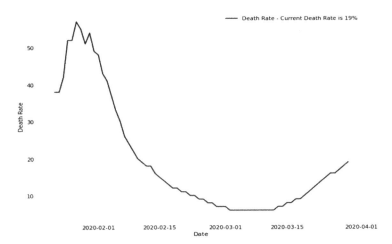

FIGURE 5.4   Death rate vs. date.

Then we used FB Prophet for forecasting the total number of cases Worldwide for the next 14 days till 12th April 2020, as the data present was till 29th March 2020. For growth forecasting, it is necessary to know whether the number to be predicted is expected to keep on growing without any bounds or there is a non-linear growth that saturates at a capacity [14]. In this case, the total number of infected people or confirmed cases is expected to saturate at a specific level depending upon the population. So, the maximum value of infected persons in both the training dataset and the future dataset is set as 1.5 million, and logistic growth is used for the prediction as logistic growth is general when the growth cannot exceed a certain maximum [14]. The logistic model in its basic form is represented as given in Eqn. (4) [14].

$$g(t) = C/(1 + \exp(-k(t-m)))\qquad(4)$$

where; C is the carrying capacity; k is the growth rate; m is an offset parameter.

The daily_seasonality, weekly_seasonality as well as annual_season-ality has been set to True for the forecasting. The future prediction is shown in Figure 5.5.

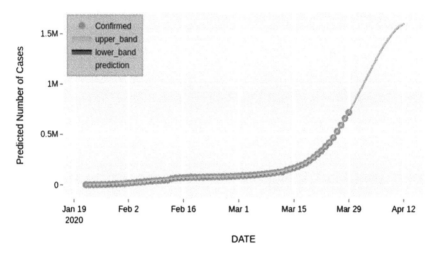

FIGURE 5.5   Prediction of confirmed cases using time series analysis.

It is seen that the upper limit is around 1.529 million while the lower limit is 1.524 million. The comparison between the predicted output and the actual number of confirmed cases until 3rd April 2020 has been done for testing the accuracy of the model as shown in Figure 5.5. We see that the predicted output is almost equal to the actual number of confirmed cases in the world and the mean percentage error is equal to 1.082%.

In the next part of our work, a comparison of the death rate, confirmed, and recovered cases for Italy and China is done. This study is done to see how different preventive measures can help in flattening the curve of confirmed cases and decreasing the death rate. We see that the confirmed cases in Italy are still increasing exponentially while the cases in China have started to flatten. It is also visible that in China, most of the cases have concluded as the line of recovered cases is moving towards the total cases as shown in Figure 5.6.

On forecasting the future values for China, it is observed that the graph remains flat and the prediction at the end of the 14-day forecast is nearly 82.90 k while the upper limit of the forecast is 86.89 k while in the case of Italy the forecast shows an increasing trend and at the end of the forecast period the prediction is 134.86 k and the upper limit is nearly 135.65 k as shown in Figure 5.7.

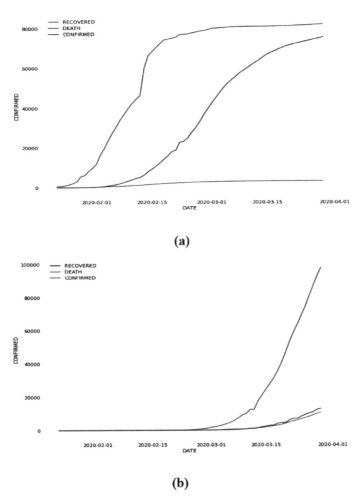

(a)

(b)

FIGURE 5.6   (a) Confirmed vs. death vs. recovered in China; (b) confirmed vs. death vs. recovered in Italy.

Figure 5.8 shows the comparison between the death rates of the two nations; it is seen that the death rate in China has decreased exponentially to 4% from a maximum of 60%, while the death rate in Italy is still high at 45%. This suggests that most of the confirmed cases in China have recovered, while in Italy, a high part of the confirmed cases remains still under treatment while the cases which have been concluded have a large number of cases which have already resulted in a fatality.

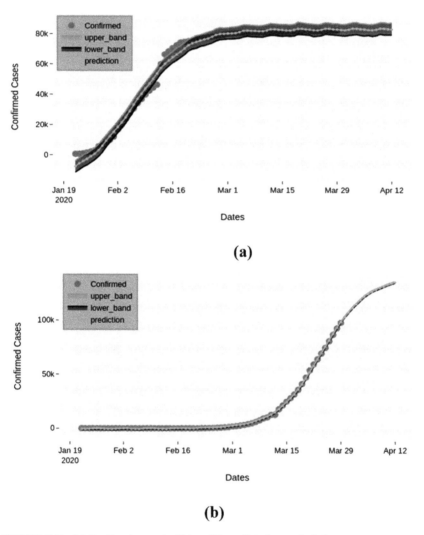

(a)

(b)

FIGURE 5.7   (a) Predicted cases in China; (b) predicted cases in Italy.

It can be seen how the difference in policies and preventive measures used in China and Italy have resulted in a difference in the amount of spread of the infection in the respective countries. In Italy during the early stages of the pandemic were met skeptically by both the general public as well the policymakers [15] which resulted in a high increase in the cases later while in China extreme lockdown measures during the early part of the lockdown [16] and other measures helped in controlling the transmission of the virus and flattening the curve.

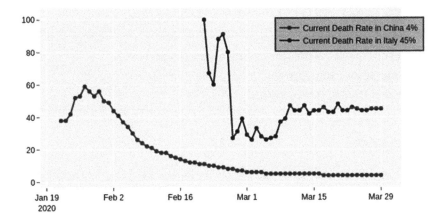

FIGURE 5.8   Death rate (percentage) vs. date in China and Italy.

5.5   CONCLUSION

We have presented a mathematical model for forecasting the number of cases using Time-Series Analysis. The forecast shows that the number of confirmed cases continues to grow exponentially around the globe. The results of this growth saturated Prophet model are commendable as the error in the prediction is as low as 1.082%. The importance of early administration of preventive measures on flattening the growth curve and lowering the death rate is also seen in the comparative study of coronavirus (CoVs) in China and Italy. The importance of implementation of preventive measures for depreciating the graph of new cases is been well compared in this chapter. Taking strict actions would not only lead to a decrease in new cases but also be fruitful in helping recover those who are infected. We have seen how decreasing the count of new cases eventually leads to the recovery of the old faster and smoother. So, to get out of this pandemic studying these statistics make us aware of the apocalypse we are heading to and how we can prevent it.

KEYWORDS

- **coronavirus disease 2019**
- **FB prophet**
- **forecasting**
- **novel coronavirus**
- **severe acute respiratory syndrome coronavirus 2**
- **time-series analysis**
- **World Health Organization**

REFERENCES

1. Sk Shahid, N., Indrajit, G., & Joydev, C., (2020). *Short-Term Predictions and Prevention Strategies for COVID-2019: A Model-Based Study*. AERU, Indian Statistical Institute, West Bengal, India.
2. Chaolin, H., Yeming, W., Xingwang, L., Lili, R., Jianping, Z., Yi, H., Li, Z., et al., (2020). Clinical features of patients infected with 2019 novel coronavirus in Wuhan, China. *The Lancet, 395*(10223), 497–506.
3. Fuk-Woo, C. J., Shuofeng, Y., Kin-Hang, K., Kai-Wang, T. K., Hin, C., Jin, Y., Fanfan, X., et al., (2020). A familial cluster of pneumonia associated with the 2019 novel coronavirus indicating person-to-person transmission: A study of a family cluster. *The Lancet, 395*(10223), 514–523.
4. Available online: Data Source-https://github.com/CSSEGISandData/COVID-19 (accessed on 22 June 2021).
5. Available online: https://www.statistics.su.se/english/research/time-series-analysis/a-brief-history-of-time-series-analysis-1.259451 (accessed on 22 June 2021).
6. Paul, N., (1975). The principles of the box-Jenkins approach. *Journal of the Operational Research Society, 26*(2), 397–412. doi: 10.1057/jors.1975.88.
7. Zhang, G. P., (2003). Time series forecasting using a hybrid ARIMA and neural network model. *Neurocomputing, 50*, 159–175.
8. Available online: https://mlcourse.ai/articles/topic9-part2-prophet (accessed on 22 June 2021).
9. Zhonghua, L., Xing, B., & Xue, Z. Z., (2020). The epidemiological characteristics of an outbreak of 2019 novel coronavirus diseases (COVID-19) in China. *Novel, C.P.E.R.E., 41*(2), 145.
10. Ratnadip, A., & Agrawal, R. K., (2013). *An Introductory Study on Time Series Modeling and Forecasting* (p. 9). Chapter 1.

11. Raicharoen, T., Lursinsap, C., & Sanguanbhoki, P., (2003). Application of critical support vector machine to time series prediction. *ISCAS '03, 5*, V-741–V-744.
12. Ferenti, T., (2017). Biomedical applications of time series analysis. *2017 IEEE 30th Neumann Colloquium (NC)*. doi: 10.1109/nc.2017.8263256.
13. Bloomfield, P., (2004). *Fourier Analysis of Time Series: An Introduction.* John Wiley and Sons.
14. Taylor, S. J., & Letham, B., (2017). Forecasting at scale. *Peer J. Preprints, 5*, e3190v2. https://doi.org/10.7287/peerj.preprints.3190v2.
15. Gary, P. P., Raffaella, S., & Michele, Z., (2020). *Lessons from Italy's Response to Coronavirus.* https://hbr.org/2020/03/lessons-from-italys-response-to-coronavirus (accessed on 22 June 2021).
16. David, C., (2020). *What China's Coronavirus Response can Teach the Rest of the World.* https://www.nature.com/articles/d41586-020-00741-x (accessed on 22 June 2021).
17. Available online: https://www.worldometers.info/coronavirus (accessed on 22 June 2021).

CHAPTER 6

Platform-Driven Pandemic Management

JAYACHANDRAN KIZHAKOOT RAMACHANDRAN and
PUNEET SACHDEVA

*HCL Technologies Ltd. Noida, Uttar Pradesh, India,
E-mails: jayachandran.ki@hcl.com (J. K. Ramachandran),
sachdeva-p@hcl.com (P. Sachdeva)*

ABSTRACT

With the unprecedented impact of the ongoing coronavirus outbreak, a large part of the medical research community believes that it is going to prolong for some time until treatment and vaccine are invented. Apart from medical intervention, there are lot of measures being taken to contain the spread of the disease. We believe the level of COVID-19 outbreak could be contained and localized, provided we apply the learnings and proven practices using technology driven solutions. The platform driven pandemic management, for public health and enterprises can act as a backbone for enabling use cases such as-location history management, contact tracing, mask, and social distancing compliance, bot driven automation and more. Reducing human touchpoints through automation being the key feature of the platform. Such platforms can be implemented on-premise or on cloud, backed by advanced application of artificial intelligence, machine learning, deep learning, natural language processing, and graph analytics.

6.1 INTRODUCTION

"No one can whistle a symphony. It takes a whole orchestra to play it."

—H. E. Luccock

What is a platform? As per the Dictionary of Computing [1], a platform is a blend of certain hardware and software coming together to support a particular activity.

In other words, a platform acts as an infrastructure for multiple applications to be developed upon, also supports multiple workflows. It is more than just a collection of applications; it allows each of those applications to work in tandem, thereby enabling a unified experience to meet the goal.

In this chapter, we are going to do a dive into the world of utility platforms and their relevance to pandemic management. We will cover use cases for public health and enterprises, along with various technologies that are crucial for powering the platform. As data management is usually at the center of such implementations, hence we will look at security and data privacy dimension as well.

First, let us understand the impact of the pandemic and what kind of solutions are most prevalent at present.

6.2 IMPACT OF PANDEMIC AND UPCOMING SOLUTIONS

As per WHO timeline, "On 31st December 2019, Wuhan Municipal Health Commission informed about acute cases of pneumonia in Wuhan, Hubei province." In the next two weeks, the first case was recorded out China in Thailand. In the next two months, the virus made its way to countries around the world, and on 11th March 2020, WHO (World Health Organization) characterized it as a pandemic [2].

The infection has grown exponentially with confirmed cases from ~4 million on 7th May 2020 to ~26 million as on 3rd September 2020. The risk to human lives is more than ever, with confirmed deaths standing at ~850,000 [3], leaving the US (United States), Europe, Brazil, and India the worst hit.

An estimated 40% of the world population experienced a total lockdown. The economic impact was unprecedented. A loss of US$ 4.5 billion per day was estimated during the first three weeks of the lockdown [4, 5]. International Monetary Fund (IMF) estimates more than $9 Trillion as loss to the global economy.

In response to contain the pandemic, stress was added on social distancing, testing, and tracking, travel restrictions and use of PPE with special focus on personal hygiene. Physical separation measures have had a significant impact on reducing transmission of COVID-19. There is a correlation between mobility reduction during State/Countrywide lock-down and reduction of transmission. Early testing enabled to minimize the spread. Contact tracing for subjects who are assessed positive is a key lever being used across regions. Some regions saw a resurgence of cases driven by imported cases. A variety of measures to spread, including contact tracing and self-isolation or imposed quarantine are applied. In terms of personal protection equipment and hygiene, early evidence in studies suggest that surgical masks catch both large and small droplets, suggesting the use of masks as a means to control the spread of the virus. Also, cleaning of surfaces, disinfection, and regular sanitization considered a best practice.

This period also saw a sudden increase in technical solutions in the areas of wearables, contact tracing, and utility platforms. For example, A home quarantine tracker-wearable wrist band [6], which can help in monitoring a group of people for them to be traced, held indoors, and prepare for social isolation. The applications are immense, consider for instance, asymptomatic person can be at once tagged at a place of mass movement and can also be enforced upon for home quarantine tracking. The Command Center can continuously check for alarms, initiate communication for intervention, and enable reporting. Alerts can be raised when the subject moves outside the geofenced area, when the tag is disconnected or is idle, or even whenever a mobile device is powered off or Mobile data/Wi-Fi is turned off.

Another such example is a wearable thermometer for real-time temperature monitoring [7], wherein the device is capable of wirelessly delivering temperature metrics and alerts. Wearables are also making entry into the manufacturing industry as well, wherein wearables can check workplace contact between employees [8].

Contract tracing is another area which has seen unprecedented growth in terms of solutions. WHO describes contact tracing as a three-step process [9]:

- **Step 1: Contact Identification:** It begins by asking about the person's activities, about events and responsibilities of the people around them, including personal, professional, and other support personnel.
- **Step 2: Contact Listing:** It includes enlisting of all persons, who are identified to have come in touch with the subject. All contacts are supplied the relevant information about prevention of the disease, and in certain cases, quarantine or isolation is needed for persons who are at greater risk.

- **Step 3: Contact Follow-Up:** It includes regular follow-up with the subjects and checking their symptoms and tests for signs of infection.

Various countries have been making substantial provisions for contact tracing. Some of them fall back upon manual methods of contact tracing like US, According to report by Johns Hopkins Bloomberg School of Public Health and the Association of State and Territorial Health Officials, "US Congress would need to provision about US $3.6 billion in emergency funding, which includes onboarding of 100,000 personnel to perform contact tracing" [10].

This period also saw a use of technology to achieve contact tracing at scale. The graphic shared below (Figure 6.1), is a depiction of how a location driven contact tracing is performed. The contacts are identified using location data and matched with location data of other individuals using the app. Individuals who have encounter a confirmed subject are duly notified and suggested to take necessary actions, including isolation and quarantine of the identified contacts [11].

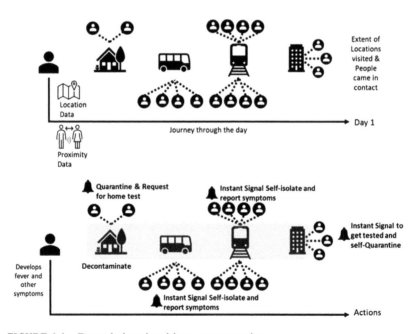

FIGURE 6.1   Example-location driven contact tracing.

Many countries adopted the use of mobile apps to track and trace citizens. Table 6.1 mentions countries and their official contact tracing app [12].

TABLE 6.1 Countries and Their Official Contact Tracing App

Country	App	Country	App
Australia	COVIDSafe	Iceland	Rakning C-19
Austria	Stopp Corona	India	Aarogya Setu
Azerbaijan	e-Tabib	Ireland	COVID Tracker Ireland
Bahrain	BeAware Bahrain	Israel	HaMagen
Bangladesh	Corona Tracer BD	Italy	Immuni
Canada	COVID Alert	Japan	COCOA-COVID-19 Contact-Confirming Application
China	App in conjunction with Alipay and WeChat	Jordan	AMAN App-Jordan
Columbia	CoronApp	Latvia	Apturi COVID
Croatia	Stop COVID-19	Malaysia	MyTrace, Gerak Malaysia, MySejahtera
Czech Republic	eRouška	Nepal	COVIRA
Denmark	Smittestop	New Zealand	NZ COVID Tracer
France	StocCovid	North Macedonia	StopKorona!
Fiji	careFIJI	Norway	Smittestopp
Finland	Koronavilkku	Qatar	Ehteraz
Germany	Corona-Warn-App	Saudi Arabia	Corona Map Saudi Arabia
Ghana	GH COVID-19 Tracker App	Singapore	TraceTogether
Gibraltar	BEAT COVID Gibraltar	Spain	Radar COVID
Hungary	VirusRadar	Switzerland	SwissCovid

With the upcoming of official contact tracing apps, concerns about data privacy also gained momentum. We will be covering security and data privacy aspects in Section 6.5. These apps used various techniques for setting up links. Prominent techniques included GPS tracking, Bluetooth, and QR codes.

Utility platforms for pandemic management also gained traction. Identigy Pandemic Response Tool [13] is meant to enable government agencies to identify possible subjects and verify if they have adequate permissions for movement and taking journey to other places. The focus point of this tool was to have a managed response with respect to journey management.

COVID-19 pandemic has been unprecedented in many ways. We have access to technology that can help us prepare a response that is integrated, scalable, and effective. Many conglomerates, service providers, and technology companies have accepted that it may take longer for us to get a cure, until then, the best bet is to use technology to safeguard people, lifestyle, and economy while dealing with the virus.

In Section 6.3, we will understand why a platform approach is the most proper step forward.

6.3   WHY PLATFORM?

"Alone we can do so little; together we can do so much."

—Helen Keller

The majority of the solutions that we discussed in the earlier Section 6.2 have limited ability to supply a complete picture. Leading to a fragmented response to the pandemic rather than a cohesive one.

It is not the first time that such a pandemic has struck. In the early 20th century, the Spanish flu had a devastating impact on the world. Due to non-availability communication and advanced technology, humanity was on its own to deal with the pandemic using traditional means. Today, we have access to the most advanced technology than ever. It is important to use the means in the best possible manner to counter the pandemic, and Platform Driven Pandemic Management is one such means. It strengthens the belief that the outbreak can be held and localized by using technology-driven solutions.

Pandemic Management solutions can be divided into two-point solutions and platform solutions. Point solutions supply best-of-breed solution for a specific use case. They majorly need the ability of a specialized technology partner and can replace specific part in the ecosystem. Nevertheless, there are certain shortcomings. The use of multiple point solutions leads to managing multiple systems, databases, and vendors. It makes user training complex, as multiple parties need to be coordinated with. It makes troubleshooting difficult due to complex ecosystem of standalone technologies. It lacks a 360° view of interactions across touchpoints which is crucial for pandemic management.

On the other hand, a digital platform as defined by Evan Bottcher "… is a foundation of self-service APIs (application programming interfaces), tools, services, knowledge, and support which are arranged as a compelling internal product. Autonomous delivery teams can make use of the platform to deliver product features at a higher pace, with reduced co-ordination" [14]. It supplies an integrated and end-to-end visibility throughout all the modules. They enable 360° view of various stakeholders, interactions, and channels. As a cohesive unit, it supplied cross-channel reporting and tracking. With single technology partner for all troubleshooting and maintenance issues it results in fewer integrations, lower cost of implementation and training costs.

Platform solutions bring synergies across tactical solutions, supply faster resolution through coordinated actions. They supply a capability to have a 360° view of important actors and activities across channels, and thereby enabling cross channel reporting and tracking.

"The whole is other than the sum of the parts."

—Kurt Koffka

In the following subsections, we will learn about how platforms are transforming public health and pandemic management in the enterprises.

6.3.1   PLATFORM FOR PUBLIC HEALTH

In September 2019, John Hopkins Center for Health Security created a report on "Preparedness for a High-Impact Respiratory Pathogen Pandemic" for global preparedness monitoring board (GPMB). Report inspects in detail the readiness for pandemics as caused by "high-impact respiratory pathogens" [15]. The report cited challenges that are posed by High-Impact Respiratory Pathogens; they are shown in the diagram below (Figure 6.2).

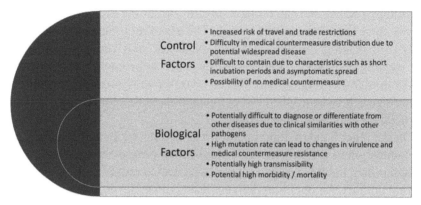

FIGURE 6.2　Challenges caused by respiratory infections.

To respond to such pandemics, investment in abilities of public health infrastructure, national, and global surveillance, frameworks for information sharing, community engagement, risk communication, rapid vaccine development, and non-pharmaceutical interventions are important.

Platforms for Pandemic Management in public health play a vital role in synchronizing activities and combining the information for the wider good. The graphic shared below (Figure 6.3) supplies a functional view of how various components can come together for pandemic management in public health setup.

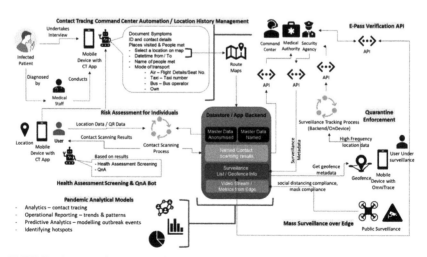

FIGURE 6.3　Functional view-platform driven pandemic management for public health.

When it comes to public health in the times of pandemic, a platform combines surveillance, monitoring, and analytics, which are the major enablers in containment and controlling of the spread. The graphic above (Figure 6.3) depict how the platform driven pandemic management uses various use cases and brings them under a single umbrella for a coordinated response.

The journey starts with an infected subject, who walks into a health facility for diagnosis by a medical staff. With the prevalent use of smartphones and the help of official apps, the health workers support the infected patient to provide details using a chatbot interface. The chatbot probes by performing a series of smart interrogations for recording symptoms, demography information, location, and other details as required for offline contact tracing. The questions related to places visited and people met are important to stitch the route maps offline [16]. Location details along with the time period during which the subject was present at the location, besides this, the bot interface gather details about individuals the subject came in contact with, and the mode of transport used to go to the place. All this information is crucial and is shared with various agencies through designated digital channels.

With the arrival of authorized contact tracing mobile applications, it is much easier to gather these details without major human inputs. Nevertheless, a sizable portion of the population might choose not to use them. In such cases, location history is set up based on such questionnaires. Once the location history is gathered from online or offline sources, the information is stored in an anonymized form, while some information must be in identifiable form with adequate security and privacy measures.

This collated location history forms the base to enable contact scanning process. People using official contact tracing app can get their individual location history compared against the anonymized location data collected in the earlier step. An ideal contact tracing takes both time and space overlap in consideration, that is the duration and proximity of the overlap. Based on the number and degree of overlaps, the individual is provided with their risk profiles and suggested the next steps. For example, in case of minimal risk, the individual can be directed to proper flow using the bot interface to get answers to often asked questions or in case of specific queries be handed over to a helpline for human aid. Similarly, in case of substantial risk, the individual details are gathered in the central store, and are marked as subjects with elevated risk. This information is shared with authorities for required intervention.

Mass surveillance is also an integral part when it comes to public health management. It allows to get insights related to compliance on the ground [17]. Compliance for keeping 6-feet social distance, mask adherence or to prevent crowding in certain places. With the help of staffed or unmanned aerial vehicles (MAV/UAV), in other words using drones, guidelines are enforced. The collected metrics make their way to the common data platform, for it to be processed and reported upon. Availability of such metrics in near-real time increases the effectiveness and efficiency for controlling the spread on the ground and to find actual or potential hotspots.

Non-pharmaceutical intervention such as Quarantine enforcement is crucial to keep infected subjects and people who meet them, under surveillance. To make it effective various agencies must come together for information sharing and implementation on the ground. There are several ways by which enforcement is done. In a more technology-oriented fashion, it starts with information exchange between various agencies. The information not only contain the information about individuals but also their real time location and area around which movement is restricted. Usually, such implementations need wearables for constant tracking, however, various other technologies such as smartphones, Bluetooth low energy (BLE), computer vision also help triangulate the subjects. It is like setting up a geofence, i.e., an alarm event is generated when the subject steps out of the allowed radius, and the agencies are duly notified.

All the crucial data elements collected in the process further enrich the data platform for various applications. One of them is e-pass verification and issuance, which is achieved as a lookup against the central data repository of known subjects, elevated risk, and quarantined subjects. Advanced analytics is applied on the data collected from the process for predictions, modeling, and operational reporting to find trends and patterns.

Cellular networks also play a role in supplying data for analytics, using the data gathered through the network providers, the researchers can analyze the activity for the various cell locations. Cellular regions with excessive concentration and movement can be deemed as regions with greater risk [18].

Platform-driven pandemic management improves the usefulness of these point solutions by a magnitude. In one such attempt Government of India, is promoting the Aarogya Setu app, which sits at the center of India's response to the pandemic. The app has proved to be a single window for the government to share guidelines and enable citizens with various useful features, including the latest updates, contact tracing, checking risk profile of family members, and it is expected to be an important enabler to support vaccination drives as well [19]. Some of the prominent features include scanning of nearby Aarogya Setu users using BLE scanner, sharing of tokens with nearby user over BLE, sharing of location-based statistics and national statistics, a self-assessment wizard as per ICMR guidelines. Also, information about emergency helpline details, approved labs, and e-pass facility. Aarogya Setu app is at the center of public health and pandemic management in India and has been instrumental in identifying clusters and hotspots.

6.3.1.1   PLATFORM FOR TELEHEALTH OR TELEMEDICINE

There are other crucial elements today when it comes to public health. Due to pandemic, the medical system faces a lot of pressure in terms of an increase in the number of critical patients, leading to delays in getting treatment of other ailments and crucial interventions like surgeries. Tele-health or Telemedicine have gained popularity amid this chaos. As per AAFP, Telemedicine is an enabler to provide medical care from distant counties, cities, or even countries [20].

Government of India has launched the National Teleconsultation Service which provides healthcare services to patients in their homes with the help of video-based clinical consultations with a doctor in a hospital. The platform has gained acceptance with more than 18000 consultation hours. The graphic below (Figure 6.4) shows the flow for eSanjeevaniOPD-National Teleconsultation Service.

In US, close to half (48%) of all US. Physicians are now treating patients through telemedicine, up from 18% in 2018, according to new data from Merritt Hawkins [21]. Teladoc Health, is one of the biggest software suppliers in this area, reports 2.8 million virtual appointments in the second quarter of 2020, which is three times compared to 2019 [22].

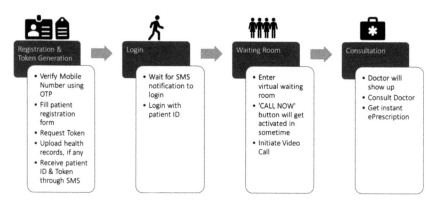

FIGURE 6.4   Flow for eSanjeevaniOPD-National Teleconsultation Service.

6.3.2   SUPPLY CHAIN MANAGEMENT PLATFORM

Supply chains witnessed a massive surprise with the advent of the pandemic. The chaotic situation led to protectionism and countries started banning exports, hoarding essential medical supplies, ordering larger volumes, etc. A similar phenomenon was felt across the entire supply chain of manufacturers, logistics providers, distributors. None of the stakeholders had any visibility of the supply chain and how to manage the situation. The whole ecosystem was dependent on production from few countries like China and other low-cost countries. The overwhelming situation created by the pandemic challenged the fundamental principles of just in time ordering and is that the right approach for healthcare where lives of people are paramount.

We faced different challenges on the ground at hospitals. There was concern on availability of beds, personal protective equipment (PPE), Testing kits, anti-viral medicines, vitamin tablets, etc. Supply chain management of PPE and information about other medical supplies such as status of bed availability across medical centers with or without ventilators is important. As frontline staffs put their own lives in jeopardy to save patients, on the other hand, the healthcare organizations fight to safeguard them with sufficient PPE.

"What gets measured gets improved."

—Peter Drucker, considered the father of Modern Management

All these challenges enhance the need for multiple new capabilities in supply chain platforms consisting of features like scenario planning, supply chain risk assessment, supply chain visibility, automated order placement, safety stock management, supply disruption prediction, etc.

Procurement platforms that instantly links buyers with sellers of PPE thereby, streamlining the process and reduce the turnaround time. These platforms safeguard against price gouging, find non-traditional suppliers, displays inventory updates, and offer speedy connections to vendors. The platform enables commodity planning which involves identification of the types and specifications of commodities required and their prioritization. With capabilities of quantity forecasting, it calculates the quantities of key commodities that will be necessary to deliver care in case of a public health emergency. For this, it takes demographic data, geographic factors, timeframe of response, consumption rate, patterns from prior epidemics and epidemiological behavior of pathogens in consideration. Procurement and sourcing of health commodities involves obtaining essential commodities from local and international suppliers in preparation for and during an outbreak. Also, such platforms enable efficient stockpiling for ready access to the supplies.

Health care supply chains are complex in nature, with many intermediaries, the platform-based implementation can help reduce the complexity by connecting suppliers to the health care providers while taking care of crucial aspects involved in health care supply chain management.

In the following section, we will learn about how platforms are transforming pandemic management in the enterprises.

6.3.3 PLATFORM FOR ENTERPRISE

Like platforms in the public health space, platforms in enterprise space play a significant role in coordinating the actions across the various groups to localize and contain the infection. It is crucial as organizations resume activities with workforce joining back from lockdown. Enterprises have several systems in place which can be used for making the workplace safe for the employees. Besides, they are liberal in making investments that improve the safety levels. Systems such as building access management, IT infrastructure, network of CCTV surveillance pave the way for enabling capabilities of contact tracing, location data management, mass surveillance for social distancing and mask compliance in the work area. Also, many organizations regularly check with their employees to know about health and safety. Information collated from such channels allows to do a risk assessment at the organization level, thus allowing stakeholders to make informed decisions to control and safeguard against creating a chain of infection.

Individuals such as building administrators, human resource personnel, security head and department heads require the information to act upon. The information includes but not limited to current occupancy and headcount of employees in the premises, the locations that need regular sanitization, ability to find the associates who may have encounter an infected subject in case some employee turns positive. Also, to know the compliance levels for corporate and federal guidelines. The information is combined at various levels of location, department, and enterprise level.

The graphic below (Figure 6.5) shows the day-to-day office journey of employees. The employee moves reports at office entry to gain entry to the work area by showing the photo ID card. In the work area, they use the co-working spaces, connect to enterprise network using WIFI or wired LAN (local area network). Also, they make use of common spaces like lobbies, elevators, breakout areas, and cafeterias. During their stay, employees must follow corporate and federal guidelines to keep the area safe for themselves and for others.

A data trail is left in the form of access logs, network logs, and video feed from CCTV, which is used by such platforms and enable use cases to actionize the response plan. For example, Physical Identity Access Management systems that are prevalent in organizations worldwide are being used for risk scoring, as a measure for pre-emptive risk assessment before permitting access. Integration with travel and HR (human resources) systems opens the possibility to map travel information to enrich the risk profile.

There has been growing collaboration and cooperation between business organizations and government to respond to the pandemic. In the earlier Section 6.2, we talked about official contact tracing apps by various governments around the world. These apps, besides enabling the government to track and trace contacts for infected subjects, also allow individuals to provide their individual risk profiles. These risk profiles are an important piece of information for organization to know about their employees. For example, The Government of India has announced a new Open API service for the Aarogya Setu app to allow businesses and organizations to have more control over the health status of their employees [24].

FIGURE 6.5   Day-to-day office journeys of employee.

Like public health, utility of platforms for pandemic management in enterprises is equally important. As organizations need to bring millions of employees back to work, they need to provide a safe and secure environment that instills confidence in their minds [23]. Such platforms in the enterprise space are:

- A key pillar of business continuity planning;
- Essential to increase confidence in society, business owners, investors, partners, employees, and customers to ensure an effective system in place;
- Enhances risk management;
- Ensures personal safety, safeguard against reputation loss, legal, and regulatory implications. In the following section, we will look at those technological enablers for such platforms.

6.4   TECHNOLOGY SUPPORT FOR THE PLATFORM

Coronavirus 2019 outbreak and its magnitude is comparable only to the 1918 Spanish Flu. The technology support that we have today is better than it has ever been. There are certain technology elements that have been at the core of pandemic response-driven through platforms.

6.4.1 CLOUD

When we talk about cloud computing, there are certain features or characteristics that come up. With today's cloud capabilities, one can request for computer, storage, and network resources as and when required, access, and achieve high availability and serve users across the globe using dedicated networks, house multiple applications and serve users using the same underlying implementation. Expand or contract the footprint with elastic resource allocations and scale to any capacity as required [25]. Cloud computing provides the barebone infrastructure for such platforms to be built upon. On-demand resource allocations allows to provision or de-commission resources like storage space, virtual machines, database instances, etc., exclusive of any manual intervention. Broad network access ensures cloud computing resources and accessible by different platforms via a network such as the Internet, or in the case of private clouds, it may be a LAN. With the availability of a pan-geo connectivity infrastructure, the quality of the service is supported. Cloud computing resources support a multi-tenant model. With the help of multi-tenancy support multiple implementations, each serving a different customer can share the same platform or the same infrastructure, at the same time keep privacy and security over their data. It is made possible by pooling resources; it is important to ensure that resource pooling is done right, and it should not hamper the critical applications. Cloud service providers (CSP) offer a large catalog of automation capabilities for rapid provisioning and de-provisioning of resources through their data centers which are found worldwide, allowing the critical applications built and served using platforms to be scalable and be available.

6.4.2 ARTIFICIAL INTELLIGENCE AND MACHINE LEARNING

Foundations of artificial intelligence (AI) and machine learning (ML) allow us to have consistent performance which is human comparable. In the fight of pandemic, AI, and ML have been playing a significant role to find people who are at risk, screening, and diagnosing, expediting drug development, even finding existing drugs which are effective in the recovery process, and predicting the spread [26]. Based on important risk factors, it is possible to find how likely a person is to get affected by COVID-19. The factors may include age, comorbidities, personal hygiene, extent of interactions, location, etc. For example, a vulnerability index devised by DeCaprio and team [27], using ML to forecast the possibility of a subject getting clinically severe [28]. Also, using the advanced ML techniques to diagnose the infection by using facial scans and thermal screening; and using chatbots with cognitive capabilities to assist subjects based on a self-assessment. For example, a coronavirus (CoVs) "self-checker" tool has been developed by the CDC (center for disease control) and Microsoft developed, to assist subjects to decide whether they need to get hospitalized or not [29]. During the Ebola outbreak, researchers had fall back on ML to find the molecules which are active against the virus in an aim to speed up the drug development [30].

6.4.3 COMPUTER VISION

Computer vision is a multidisciplinary playing field that complies with how computers can develop a high-level understanding by interpreting digital images. It has made great progress in the past few years, due to the success of deep learning, a subfield of ML [31]. As per a survey conducted by Anwaar Ulhaq and team [31], computer vision has been playing a significant role in diagnosis, prevention, and, and control, clinical management, and treatment. For diagnosis, analysis of CT-scan and X-Ray images is being used. Similarly, for prevention and control, analysis of images acquired from thermal cameras is being chosen. In the enterprises, computer vision technology is being used to analyze CCTV feeds which currently check areas in and around the premises. With the help of computer vision, areas are watched for social distancing and mask non-compliance and raises alerts for required stakeholders to intervene, thereby making the premise safe. The benefit of using this technology is that it can perform the required task 24×7 with same service levels. The information or metrics acquired from this is used to create baselines by identify hotspots, i.e., The places where a considerable number of violations are happening.

6.4.4   MOBILE APP ADVANCEMENTS

Mobile apps have seen some major acceptance and advancements during pandemic, especially in the areas of healthcare, education, entertainment, and remote work. However, use of mobile apps for contact tracing has the most relevant when it comes to pandemic response. As mentioned in the earlier Section 6.2, many countries launched their official contact tracing apps to safeguard the interest of its citizens. They used various techniques to triangulate the location, proximity, and time overlap of the intersection between devices held by individuals. Techniques include the use of GPS data, QR codes and Bluetooth. The initiative is duly supported by Apple and Google for their respective mobile platforms, iOS and Android, to come together and build a common API for seamless information exchange between devices to enable exposure notifications. The API works using a decentralized identifier system which is based on randomly generated temporary keys localized on the user's device. Common API allows public health agencies to use it in official apps and to define potential exposure in terms of exposed time and distance. There has been concerns with respect to privacy of individuals amid the development and implementation of these techniques. We will cover them in the "security and data privacy" Section 6.5 later.

6.4.5   WEARABLES

During the CoVs pandemic, wearables have gathered universal acceptance especially in the space of health monitoring. One of the revolutionary application of wearables can be seen in Smart clothing, where built-in sensors enable remote monitoring of patients, thereby eliminating the need to visit the hospital. Respiratory monitoring is another application where sensors track breathing parameters of the subjects and relay them to health-care providers for round the clock monitoring of affected patients. There is another class of versatile wearables that capture various metrics such as heartbeat, blood pressure, blood sugar and oxygen levels and provides them to the health care providers for required intervention. Wearables are also getting prevalent in the manufacturing industry, where employees are provided with bracelets. The bracelets have the sensors that record the movement of individuals, proximity to another employee and overlap of the intersection. Some versions even vibrate to raise alarm if guidelines are breached. Certain group of population believes they are intrusive however, they are remarkably effective when it comes to gathering or actionizing in real-time.

6.4.6   *NATURAL LANGUAGE UNDERSTANDING, PROCESSING, CONVERSATIONAL AI*

Natural language processing (NLP) and understanding has been among for a while now. Its more advanced version, Conversational AI has garnered visibility. Conversational AI is a branch of AI that deals with human like interactions. For example, the voice assistants that are available in the form of Siri, Alexa, Google Assistant have their roots in Conversational AI. The applications span across the areas of health care, contact center, and chatbots. In the area of healthcare, it is used to equip patients with information about the CoVs, assess their risk by asking a series of questions to arrive at risk profile, support preventive care, and guidance to local resources. In the contact center, applications powered with NLP, NLU, and Conversational AI are ensuring to meet the unprecedented inflow of queries, enquiries, and S.O.S calls. The applications are supported by human in the loop for interventions that require handling by human agents.

The technological components discussed above-cloud computing, AI and ML, mobile app advancements, wearables, and conversational AI, though they may appear to be enabling applications that may appear to be Siloed, however, these components also have the capability to come together and complement each other. The platforms for public health and platforms for the enterprise may use them for responding to the pandemic on the ground.

6.5   SECURITY AND DATA PRIVACY

As the case is with any application which gathers, process, and store data, there are certain responsibilities that must be followed at all costs. Data privacy concerns go mainstream in the public health scenario, while for the enterprise, it is subdued since in the majority of the cases, mass surveillance is already being performed as covered under the employment agreement.

The first aspect in the data journey is of data gathering. During collection of data related to individual health, location, and movement need to be backed by an explicit Opt-In. By implementing an Opt-In, the purpose is to get a consent. This ensures that the subject is aware of the activity, especially why is there a need to collect the data, what elements are being collected, and how the data that is collected is used to achieve which goal. The other way is to allow an Opt-Out, which means, data is collected for all subjects by default until they explicitly want to opt-out of it. It has been seen that with Opt-In, there is an exceedingly high chance that subjects may not agree for it in the first place. Ultimately leading to low acceptability and limited or no data to be collected. That is why the majority of applications are implemented as a default Opt-In with no opt-out in majority of the countries, and with a choice to opt-out in countries or regions with stringent privacy laws such GDPR (general data protection regulation), CCPA (California Consumer Privacy Act), etc.

The technology used for contact tracing is also important to understand the data privacy aspects. The three major techniques include tracing based on GPS coordinates, exposure based on BLE, tracing based on QR codes. Among these, contact tracing through mobile using Bluetooth based API fairs best in terms of privacy, followed by contact tracing through mobile using QR codes, followed by contact tracing through mobile using GPS locations which fairs incredibly low on data privacy scale [32]. It all comes down to collecting that is needed with consent and using techniques that promote privacy.

The second aspect in the data journey is of Data processing. Even though care is taken at the data collection step in the form of explicit consent and collecting only data that is needed, it is also important to ensure that data collected when processed should not identify the individual person. For health care applications, it is important to store individual preferences, however, the way they are stored and processed is still crucial. It can be ensured by anonymizing the dataset to confirm that it cannot be used to directly identify the person. Also, for the data that has reference to a person's identity should be stored with added data security controls and limited access. The data scientists and the models that they build are majorly feature driven and do not require PII (personally identifiable information) for achieving goals. Similarly, for features enabled through computer visions should ensure that only metrics are being used for processing rather than finding the person behind those metrics. In this case also, even if the PII or individual details are needed for the final application, the data must be stored with data security controls and limited access. The use of advanced techniques like federated learning, it allows the algorithm to get processed where the data is, rather than keeping it at a central place [33].

This brings us to the third major aspect, which is data retention and storage. With the use of cloud computing, we can keep data of persons from a specific region within the physical boundaries of that country, provided cloud service provider has a presence. Cloud computing has also enabled distributed and decentralized processing, thereby minimizing the concerns on privacy. Having said that, it is important to use data only for its intended use, only for the duration of the intended use and purge it safely once it has served its purpose. For example, a CCTV feed after it has been analyzed and the metrics have been gathered should be discarded as it has served its purpose. Also, for serving real time use cases, we need not keep data beyond a certain period, post which it should be safely purged.

It is always a topic of debate, which has more importance, a human life, or data privacy. There is no direct answer to this. However, a platform-based approach is the only choice to have a standardized implementation that brings all relevant components together as a cohesive unit to save human life at the same time ensures proper safeguards are in place to protect privacy.

6.6 CONCLUSION

A platform approach helps to bring together diverse technologies, processes, systems, innovations and provides a common framework where there is better coordination, leverage of synergies, and a common cause for a larger purpose. Traditional ways of getting things done has been disrupted by technology platforms which had brought in multiple stakeholders on a common framework like eCommerce, social media, open education, risk management. We have seen how such platforms have broken barriers, created a level playing field and brought in unprecedented efficiencies. The future is all about collaboration and bringing the best of everyone, and it is even more important when we are challenged by an existential crisis due to the pandemic. Humanity has always produced breakthrough ideas during crisis, and it has brought the best in them. This crisis is for us to exploit, technology is going to be at the root of it, and platforms is the way to go.

"The line between disorder and order lies in logistics…"

—Sun Tzu

KEYWORDS

- **application programming interfaces**
- **artificial intelligence**
- **Bluetooth low energy**
- **cloud service providers**
- **global preparedness monitoring board**
- **local area network**
- **machine learning**

REFERENCES

1. FOLDOC: Free On-line Dictionary of Computing, (1994). *Platform* [Online]. Available: https://foldoc.org/platform (accessed on 22 June 2021).
2. WHO, (2020). *Archived: WHO Timeline - COVID-19* [Online]. Available: https://www.who.int/news-room/detail/27-04-2020-who-timeline---covid-19 (accessed on 22 June 2021).
3. WHO, (2020). *WHO Coronavirus Disease (COVID-19) Dashboard* [Online]. Available: https://covid19.who.int/ (accessed on 22 June 2021).
4. Business Line, (2020). *COVID-19 Lockdown Estimated to Cost India $4.5 Billion a Day: Acuité Ratings* [Online]. Available: https://www.thehindubusinessline.com/economy/covid-19-lockdown-estimated-to-cost-india-45-billion-a-day-acuit-ratings/article31235264.ece (accessed on 22 June 2021).
5. PTI, (2020). *Experts Peg India's Cost of Coronavirus Lockdown at USD (United States Dollars) 120 bn* [Online]. Available: https://www.thehindubusinessline.com/economy/experts-peg-indias-cost-of-coronavirus-lockdown-at-usd-120-bn/article31160115.ece (accessed on 22 June 2021).
6. NASSCOM, (2020). *Home Quarantine Tracker- Wearable Wrist Band* [Online]. Available: https://nasscom.in/home-quarantine-tracker-wearable-wrist-band (accessed on 22 June 2021).

7. Cathy, R., (2020). *Rinsfox Launches Indiegogo Campaign for Developing Wearable Thermometer for Real-Time Temperature Monitoring* [Online]. Available: https://www.wearable-technologies.com/2020/04/rinsfox-launches-indiegogo-campaign-for-developing-wearable-thermometer-for-real-time-temperature-monitoring/ (accessed on 22 June 2021).

8. Cathy, R., (2020). *Estimote Introduces Wearables to Monitor Workplace Contact Between Employees to Curb COVID-19 Outbreak* [Online]. Available: https://www.wearable-technologies.com/2020/04/estimote-introduces-wearables-to-monitor-workplace-contact-between-employees-to-curb-covid-19-outbreak/ (accessed on 22 June 2021).

9. WHO, (2017). *Contact Tracing* [Online]. Available: https://www.who.int/news-room/q-a-detail/contact-tracing (accessed on 22 June 2021).

10. Jon, S., (2020). *This Contact Tracing Technology of COVID-19 in Over a Dozen Jurisdictions Could Assist Companies as they Reopen* [Online]. Available: https://www.marketwatch.com/story/this-contact-tracing-technology-of-covid-19-in-over-a-dozen-jurisdictions-could-assist-companies-as-they-reopen-2020-04-27 (accessed on 22 June 2021).

11. Ferretti, L., Wymant, C., Kendall, M., Zhao, L., Nurtay, A., Abeler-Dörner, L., Parker, M., et al., (2020). *Quantifying SARS-CoV-2 Transmission Suggests Epidemic Control with Digital Contact Tracing* [Online]. Available: https://science.sciencemag.org/content/368/6491/eabb6936 (accessed on 22 June 2021).

12. Wikipedia, (2020). *COVID-19 Apps - List of Apps by Country* [Online]. Available: https://en.wikipedia.org/wiki/COVID-19_apps (accessed on 22 June 2021).

13. Identigy, (2020). *Pandemic Response Tool* [Online]. Available: https://identi.gy/pandemic-response-tool/ (accessed on 22 June 2021).

14. Evan, B., (2018). *What I Talk About When I Talk About Platforms* [Online]. Available: https://martinfowler.com/articles/talk-about-platforms.html (accessed on 22 June 2021).

15. Johns Hopkins Center for Health Security, (2019). *Preparedness for a High-Impact Respiratory Pathogen Pandemic* [Online]. Available: https://apps.who.int/gpmb/assets/thematic_papers/tr-6.pdf (accessed on 22 June 2021).

16. Express News Service, (2020). *Route map of Kollam's First COVID-19 Patient, Who Travelled via Thiruvananthapuram Released* [Online]. Available: https://www.newindianexpress.com/states/kerala/2020/mar/28/route-map-of-kollams-first-covid-19-patient-who-travelled-via-thiruvananthapuram-released-2122423.html (accessed on 22 June 2021).

17. Jed, P., (2020). *Drones Become Part of Local U.S. Responses to COVID-19* [Online]. Available: https://www.govtech.com/products/Drones-Become-Part-of-Local-US-Responses-to-COVID-19.html (accessed on 22 June 2021).

18. Jessica, K., (2020). *Cellular Network Data Detects Potential COVID-19 Hotspots* [Online]. Available: https://healthitanalytics.com/news/cellular-network-data-detects-potential-covid-19-hotspots (accessed on 22 June 2021).

19. OpenForge.gov.in (2020). *Aarogya Setu IOS App* [Online]. Available: https://openforge.gov.in/plugins/git/aarogyasetuos/ios-mobile-application (accessed on 22 June 2021).

20. AAFP, (2019). *What's the Difference Between Telemedicine and Telehealth?* [Online]. Available: https://www.aafp.org/news/media-center/kits/telemedicine-and-telehealth. html#:~:text=Telehealth%20is%20different%20from%20telemedicine,to%20 remote%20non%2Dclinical%20services (accessed on 22 June 2021).

21. Merritt Hawkins, (2020). *Survey: Physician Practice Patterns Changing as a Result of COVID-19* [Online]. Available: https://www.merritthawkins.com/news-and-insights/media-room/press/-Physician-Practice-Patterns-Changing-as-a-Result-of-COVID-19/ (accessed on 22 June 2021).

22. Teladoc Health, (2020). *Teladoc Health Reports Second-Quarter 2020 Results* [Online]. Available: https://teladochealth.com/newsroom/press/release/second-quarter-2020-results/ (accessed on 22 June 2021).

23. Jayachandran, K., (2020). *Break the Chain of COVID-19 with Pandemic Management Platform* [Online]. Available: https://www.hcltech.com/blogs/break-chain-covid-19-pandemic-management-platform (accessed on 22 June 2021).

24. Times of India, (2020). *Businesses and Organizations can Check Aarogya Setu Status of Employees* [Online]. Available: https://timesofindia.indiatimes.com/gadgets-news/businesses-and-organisations-can-check-aarogya-setu-status-of-employees/articleshow/77702571.cms (accessed on 22 June 2021).

25. Wikipedia, (2020). *Cloud Computing* [Online]. Available: https://en.wikipedia.org/wiki/Cloud_computing (accessed on 22 June 2021).

26. Astha, O., (2020). *How Artificial Intelligence is Helping to Fight Against Coronavirus in India?* [Online]. Available: https://www.analyticsinsight.net/artificial-intelligence-helping-fight-coronavirus-india/ (accessed on 22 June 2021).

27. DeCaprio, D., Gartner, J., McCall, C. J., Burgess, T., Garcia, K., Kothari, S., & Sayed, S., (2020). *Building a COVID-19 Vulnerability Index* [Online]. Available: https://arxiv.org/pdf/2003.07347.pdf (accessed on 22 June 2021).

28. Jiang, X., Coffee, M., Bari, A., Wang, J., Jiang, X., Huang, J., Shi, J., et al., (2020). *Towards an Artificial Intelligence Framework for Data-Driven Prediction of Coronavirus Clinical Severity* [Online]. Available: https://www.techscience.com/cmc/v63n1/38464 (accessed on 22 June 2021).

29. Tyler, S., (2020). *Microsoft and the CDC Have Built an AI-Powered Coronavirus Chatbot that can Help You Figure Out Whether You Need to go to the Hospital* [Online]. Available: https://www.businessinsider.in/science/news/microsoft-and-the-cdc-have-built-an-ai-powered-coronavirus-chatbot-that-can-help-you-figure-out-whether-you-need-to-go-to-the-hospital/articleshow/74783019.cms (accessed on 22 June 2021).

30. Ekins, S., Freundlich, J. S., Clark, A. M., Anantpadma, M., Davey, R. A., & Madrid, P., (2016). *Machine Learning Models Identify Molecules Active Against the Ebola Virus In Vitro* [Online]. Available: https://www.ncbi.nlm.nih.gov/pmc/articles/PMC4706063.2/ (accessed on 22 June 2021).

31. Ulhaq, A., Khan, A., Gomes, D., & Paul, M., (2020). *Computer Vision for COVID-19 Control: A Survey* [Online. Available: https://arxiv.org/pdf/2004.09420.pdf (accessed on 22 June 2021).

32. Tomas, P., (2020). *Coronavirus: How to Do Testing and Contact Tracing* [Online]. Available: https://medium.com/@tomaspueyo/coronavirus-how-to-do-testing-and-contact-tracing-bde85b64072e (accessed on 22 June 2021).

33. Marielle, S. G., & Robert, C. M. Jr., (2020). *Federated Learning: Collaboration Without Compromise for Health Care Research* [Online]. Available: https://www.statnews.com/2020/02/13/federated-learning-safer-collaboration-health-research/ (accessed on 22 June 2021).

CHAPTER 7

Smart IoT Techniques to Improve Pandemic Outreach

D. KARTHIKA

Assistant Professor, Department of Computer Science, Chrompet, Chennai, Tamil Nadu, India, Email: Karthika.d@sdnbvc.edu.in

ABSTRACT

The COVID-19 pandemic has crossed the provincial, radical, philosophical, spiritual, educational, and pedagogical limits of the latest global problem. The healthcare infrastructure allowed by the internet of things (IoT) is beneficial using an integrated network for careful surveillance of COVID-19 patients. This device aims to improve the comfort of patients and decrease the hospital readmission rate. Implementation of the IoT influences reducing healthcare costs and improving the affected patient's medical outcome. Therefore, by presenting a theoretical framework to solve the COVID-19 pandemic, this new study-based analysis aims to investigate, analyze, and illustrate the overall implications of the well-proven IoT theory. Finally, it describes and addresses twelve essential IoT implementations. Ultimately, scholars, educators, and scientists have been compelled to suggest some productive ways to solve or cope with this pandemic.

7.1 INTRODUCTION

IoT providers have recently grown into many industries such as agriculture and the industry due to increased efficiency from the use of IO devices, intelligent watches, intelligent instruments, surveillance cameras, different types of sensors, etc. Moreover, we foresee 5G-based IoT systems and devices to expand through various fields such as vehicles, remote control of heavy equipment, ultra-high-definition video streaming or tele pressure at snow sports competitions, etc., with the appearance of next-generation 5G-communications with a higher speed and high capability. The internet of things (IoT) is a well-developed system of electronic, wireless, and mechanical interconnected devices that can transfer data across the established networks without human intervention at all levels [1]. Their special I'd numbers or codes are also connected to these products. IoT is now well established and tested technology that serves as a junction to countless strategies, immediate study, machine learning (ML) theory, sensory learning items and so forth. In addition, IoT is acknowledged in traditional everyday working as the functionality of the goods or technologies meet people's real-life requirements in various ways such as home protection, smart lights, and many other systems, which are easily regulated by mobile layers, smartphones, etc., by our daily usage. Both nations, including India, are battling COVID-19 in the current pandemic and are trying to explore a realistic and cost-efficient solution to different [2]. Physical science and engineering researchers are seeking to solve these difficulties, create new ideas, explain new study questions, create user-centric solutions, and build up ourselves and the civil society. This brief analysis aimed at sensitizing the new technologies and its relevant implications for the pandemic of COVID-19. No one predicted COVID-19 to give away as much as it did from the global economy. Technology has proved to be one of the best partners any company can have during these tough times. And now is the time to take full advantage of it. Today, I want to explain how the IoT will help any organization in the pandemic.

7.2   MAINTAINING BUSINESS CONTINUITY

Faced with this problem, the bulk of businesses have shot as they arrive. The short-term target of nearly any company out there is sustaining its activities (or, more frequently, mitigating economic damage). In order to do this, though, it is important to ensure people's welfare.

Although distance work is the best choice today, it was not feasible for many companies to move an entire workforce away from their premises overnight. This process takes time, especially if there is no previous experience. IoT may here be the secret to a smooth transition to certain changes which we face soon [3]:

1. **Digitized Processes:** This is the first step of remote cooperation facilitation. Indirect and direct operations via cloud storage can be supported by iota technologies. This allows workers to increasingly shift from on-premises activities to the digital environment. The whole team will soon be able to track the performance of these IoT tools and speak over video calls about action plans.

2. **Eliminate Contact:** If workers on the property are entirely irreplaceable, physical separation is a priority. A IoT network can optimize essential danger points, such as adjustments to transfer, overly staffed areas, and manufacturing enclosures with the approval of its workers and under local regulations. This surveillance tools offer real-time analyzes that allow organizations to make decisions based on results.

3. **Remote Management:** IoT may be used for remote surveillance and control of systems where regulatory laws absolutely prohibit operation on-site. Through integrating essential properties with cloud-based management tools, workers can access these instruments from their own homes and retain their settings over time [4].

7.2.1 DEVELOPING FLEXIBILITY AND RESILIENCE

Uncertainty is unavoidable on the route ahead. In the next few months, we are yet to explore the latest threats. We are all going to exist in a brand-new world until the pandemic stops. Relying on IoT technologies will help businesses improve unpredictable transition stability and resilience [5]:

1. **Stable IT Infrastructure:** The two cornerstones of corporate IT are cybersecurity and networking. They are more critical than ever in a global remote work scenario. One of the strongest means to create good security to malicious threats while preserving

successful workflows is a local IoT network that links people with machines and devices.

2. **Cost Savings:** IoT technology will use predictive analytics to define critical points to drive mid-term cost changes. That includes employees and machinery productivity, availability, consistency, and performance. An inventory, output capability, and supply chain administration information are also accessible in real-time through a network of sensors.

3. **Predictive Income**: Unanticipated downturns are a reality. Companies may use IoT to analyze local and consumer data and to define the right steps based on factors to sustain revenue and customer loyalty. In a certain amount of time, IoT will also take stock levels and production plans into account in order to suggest the best pricing.

7.2.2   BUILDING A LONG-TERM COMPETITIVE ADVANTAGE

Whatever market tactics we are now adopting will propel the world of post-coronavirus. The move to contactless supply and personalization of service is already clear. However, it is to emerge as powerful as possible as the vital component of this crisis. The only way to do so is by modifying this scale over time now [6]:

1. **Intelligent Acceptance:** Intelligent systems are the future of home and industrial appliances. This was accelerated further by the pandemic. Products and services to be popular soon are those which make it easier to communicate, personalize, and run.

2. **Alignment of Supply Chain:** Sharing is a huge obstacle for players in the Supply Chain. It is thankfully the key power of IoT, too. IoT is the ideal approach for reducing excessive costs, removing typical supply line risks, and exposing secret perspectives that optimize value creation, ranging from scheduling of manufacturing schedules to quality control of inventories.

3. **General Processing Method:** An IoT network can process information from almost all business processes or business protocols through ML. Combined with development programs and predictive modeling, businesses may use IoT to plan rapidly for

improvements to the industry, for example, increases in demand or supply interruptions.

7.2.3   LEVERAGE THE POWER OF IOT

In the corporate world, the IoT is a big change and will only expand over the years. Provided that the IT sector is one of the few industry sectors to have retained its business as normal, many do have the potential to externalize Java development to create IoT applications [7].

Now that the pandemic has begun several months ago, conditions are beginning to improve. This is the ability to make long-term structural reforms. Operating models have disrupted the status quo. It is time for technologies to be leveraged and long-term solutions to bring progress today and tomorrow.

7.2.4   A CONNECTED RESPONSE

How will an IoT-linked universe overcome these challenges?

1. IoT sensors may help detect suspected virus signs. Through collecting data at micro- and macro-level, policymakers may detect irregular patterns prior to being a concern. This can range from clever security cameras to gadgets such as a clever ring that can detect COVID-19 symptoms from a person to high tempera-ture patients. Cloud computing enables data to be analyzed and analyzed easily on distributed computers around the world and then made available to scientists around the globe, detecting trends and creating information for faster combating the epidemic [8].

2. Many vital infrastructures across the globe remain unconnected or do not have remote diagnostics and status alerts, much less the opportunity to solve these problems remotely. It is critical for the need for an engineer on-site to be reduced to minimal in a pandemic where physical interaction must be restricted and work-forces reduced. Full IoT linked devices allow fewer visits and a greater capacity to remotely fix technological challenges, improve up-time and shield workers from dangerous site visits.

3. The recession of COVID-19 pushes companies to innovate at an unimaginable pace. Retailers and other non-essential firms dependent on consumers visiting their shops or distributors will need to amend their market plans to raise profits. IoT can assist with this in different ways: Smart lockers and creative approaches for e-commerce operations, for instance, will minimize the strain of satisfying increased demand and include the "zero-touch solution" that makes handling lock-down environments simpler and safer for companies. In essential stores such as corner shops, supermarket shops, and gas plants, smart sales equipment can be mounted at food and beverage enterprises. The linking of these devices to IoT offers more business information, live tracking (and warnings to modify consumables to improve service levels) and marketing ads to devices [9].

4. During a pandemic, health systems are obviously burdened because of their scarce funding. In different forms, IoT can help here. For example, remote patient surveillance helps doctors to monitor patients from their homes. One such breakthrough is Philips Motiva, an immersive application for telehealth intended to track people with chronic illnesses from their own homes. This frees up beds and eliminates the danger of contracting the infection, but still makes sure that they can be easily diagnosed and taken to hospital if their health deteriorates.

7.3   HOW IOT CAN HELP CONTROL PANDEMIC?

One of the United Nations' goals for global growth is its ability to cope with infectious disease spread. This purpose was never as important during the COVID-19 pandemic. Science plays a crucial role in the combat against infectious diseases, and IoT is a critical part of our scientific arsenals. Decreased costs, autonomous, and online treatment for and diagnosis, and newly open patient details are only a couple of the forms IoT healthcare process's function. In terms of disease prevention [8–10], wireless, and IoT systems are useful methods.

We still consider IoT as a network of microcontrollers and wireless sensors, but that is just the physical IoT layer. IoT is typically spread massively in machines. The IoT's almost 24 billion intelligent computers measure and perform incomparable amounts of data. Further specifics specifically mean stronger decision-making and stronger pandemic response preparation. Both are important for disease transmission prevention and control.

7.3.1   TRACING OF TOUCH

Locate and recognize anyone who might interact with the patient at the outset of the pandemic is the most critical task. The way to avoid the outbreak. This is the way. Interviews and queries are the foundation for traditional touch tracing techniques. This is an effective, time-consuming, error-producing method. Populations through highly populated cities worsen the dilemma and illustrate the failures of the traditional method [11].

In the form of contemporary contract surveillance alternatives for precise tracking in the region, wireless technology (RFID, Bluetooth low energy (BLE), GPS, Wi-Fi, and Magnetic field signature) is the foundation. In comparison to the traditional approach, wireless technology offers knowledge of contact duration and position of checked instances. BLE will have fair accuracy location surveillance as one of the highest priorities of IoT. In comparison to Wi-Fi and cellular position, Bluetooth LE has greater sensitivity than proximity detection. This improved precision is important for classifying the contacts traced and for prioritizing the answer to close connections. For locations, like RSSI and Arriving angle (Aloe), Bluetooth LE offers a large selection of monitoring instruments. Equally popular with the Bluetooth LE standard are our smartphones and other wired wearables.

The usage of Bluetooth tags at the onset of a pandemic often strengthens the response plan. Hundreds or even thousands of Bluetooth tags and smart phones communicate in populated urban area. In order to contextualize this. In order to overcome the message crashes that may indicate the joining to the highly sensitive topic was not registered, this Bluetooth map must be configured in the system [12].

7.3.2   CHECKS FOR BIOSENSORS AND POPS

Point-of-care monitoring is another important activity for pandemic response. When we look at COVID-19 results, the lack of readily available test kits for COVID-19 has made us feel at the tip of the iceberg. If test kits were more readily available, the unregulated dissemination of COVID-19 virus in New York City could be reduced. In rural, emerging areas of the world, which lack qualified staff and fitted health facilities, cost-effective, and easy to install diagnostic devices may also contribute to an uncontrollable spread.

Diagnostic instruments need to be accurate, responsive, compact, and user friendly, aside from being economically viable. Furthermore, a minimal shape element is desired whether it can be completely or partly disposable, easy to replicate. This bill suits well for cloud-based biosensors [13]. Researchers at the Imperial College London have shown a laboratory able to diagnose diseases early on. The patient perspective is clear. A sample is removed in a unit from the patient. The measurements are done within 30 minutes.

Technically, the cartridge is loaded into the chip containing an ISFET sensor array mounted on a CMOS chip linked to the microcontroller, which transfers the data through Bluetooth on a server or a mobile application. ISFET is a biological sensor close to the transistor MOSFET, but a metal gate just replaces an ion-sensitive frame. Solution ion concentrations can be determined by the ISFET biosensor (Figure 7.1). The sensors perform reaction ion-imaging on the chip surface to allow DNA amplification to be monitored in real-time [14]. This monitoring is like the detection of viral infections. This illustration illustrates how IoT will use trivial technologies to enable computing during and to enable unprecedented treatment experiments in the field of treatment. This Lab-on-Chip technology, even though somewhat cheaper, is just like the Abbott Laboratory in a kit testing kit that has been approved and used by the FDA over the last few weeks. And in context, to bring it.

7.4 INTERNET OF MEDICAL THINGS (IOMT)

In American hospitals alone, the CDC reports the number of diseases acquired in the health system to be 1.7 million annually (the percentage of individuals affected during health visits). This number does not only rise throughout pandemics, it also leads health workers to be inadequate when we really need it. How can we reduce the risk during a pandemic of hospital infections [15]?

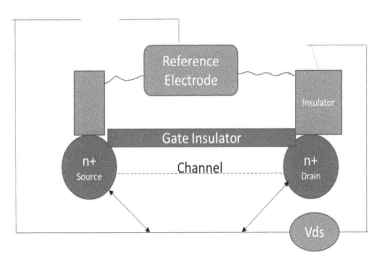

FIGURE 7.1   An ISFET biosensor.

The medical services internet (MSI) is an enticing alternative. More than 70% of healthcare professionals have already evaluated the usage of IoMT, which presents good news for future pandemic prevention. Every entity becomes a database, which is the central feature of IoT. The "object" for IoMT, whether a heart rate sensor, a wheelchair, or a portable instrument, is medical. A constant stream of patient health data (PGHD) will measure the state of the patient. A broader network can be used for the collection of data from patient populations for speeding medical science and development.

Deploying more robotics and technologies in hospitals like IoMT means less contact with infected patients and greater security for workers. Ready patient information decreases hospital visits' criteria and length. This is in accordance with the advancement of virtual visit, remote diagnostic, and tracking technology. Although IoMT and automation are and, except for pandemics, it is important to remember that technology does not eliminate the interconnection between people, which is a key component of health care. In other words, IoMT allows more time for clinicians, through patient and family consultations, to concentrate on the human side of their work [16].

IoMT also increases aged treatment or people with underlying illnesses and might result in a drastic decline in exposure to the most vulnerable groups in situations like the latest COVID-19 pandemic. During a pandemic, treatment for the least interactive elderly is crucial to protect their lives from being risky. We will sustain our elderly communities physically and psychologically healthy with automated assistants, medical devices, and intelligent residences.

7.4.1   SECURITY THREATS

Without talking about protection and privacy problems, IoMT, and IoT discussions are never complete. Although IoT technology has made sufficient progress in transferring data from cloud items, device, and data security remain a problem. Therefore, medical professionals utilize IoT in their back-end software and try to mobilize customer experience at the new frontier of IoMT [17]. Patients are genuinely scared of an innovative wireless fitness tracker that transmits vital health details constantly. In addition, in the case of a pandemic and where contact tracing is involved, data security becomes a critical issue. Who is liable for the data obtained by using Bluetooth tags in the public to enable pandemic contact monitoring? And how frequently will this sensitive material be interpreted and exploited?

In order to access health users' paws, privacy, and IoT protective vulnerabilities have to be discussed. This needs to be overcome through the collective working of regulatory, tax, medical, and technical players in the industry. Technically speaking, the security of hardware and software systems now progresses considerably [18]. For example, Silicon Labs Safe Vault technologies generate a unique signature like a birth certificate for each wireless chip. This indicates just IoT service providers and no hackers across the chip measurements made on this chip. However, the way businesses handle their personal details remains accessible to developing consumer trust.

In order to combat and monitor present and possible pandemics, IoT technologies should be a pioneer. The Mass IoT is an immense set of pandemics data and analysis [19] for mankind. It becomes more effective to track, test, and manage whole communities with IoT technologies by regulating the transmission of a disease.

7.4.2   NEED FOR THE STUDY

The global pandemic epidemic is creating an elevated number of untreated patients worldwide day by day, and the well-integrated services delivered by the IoT methodology needs to be used tremendously. In the area where the internet of healthcare things (IoHT) or the internet of medical things (IoMT) are concerned with these problems, IoT has often been used to support the desired ends [20]. The number of cases settled can also be increased and improved by observing the instructions and facilities of IoHT/IoMT.

7.4.3   KEY METRICS OF IOT FOR PANDEMIC

IoT is an advanced device that guarantees isolation for all contaminated people. It is beneficial for an appropriate tracking system during quarantine [21]. The Internet-based network is conveniently used to monitor all high-risk patients. It is used to calculate biometrically such as blood pressure, cardiac rate, and glucose content. The essential IoT advantages for COVID-19 as seen in Figure 7.2.

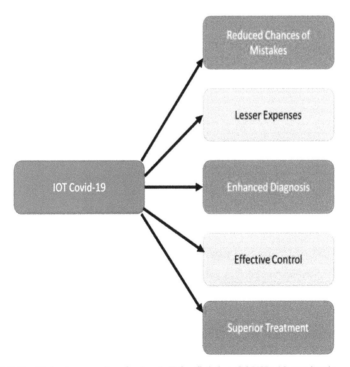

FIGURE 7.2   Major key merits of using IoT for fighting COVID-19 pandemic.

With the efficient introduction of this equipment, the productivity of medical professionals is increased and their workload minimized. The same holds true for the COVID-19 pandemic with reduced costs and defects.

7.4.4   PROCESS INVOLVED IN COVID-19

IoT is a highly advanced tool for addressing the pandemic COVID-19, which will pose major shutdown obstacles [22]. This device aims to monitor the infected patient's real-time data and other necessary data. Figure 7.3 displays the main COVID-19 IoT processes.

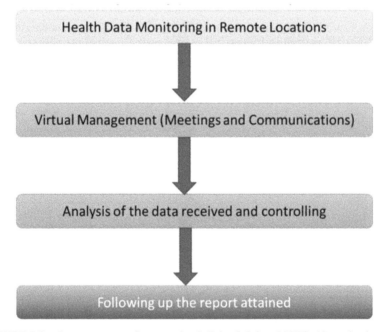

FIGURE 7.3   Step-up process for executing IoT for fighting COVID-19 pandemic.

In the first stage, IoT is used to gather health data from multiple places in the infected patient and to use the interactive management system to handle all of them. This technology allows data to be monitored and reports to be followed up [23].

7.5   THE OVERALL IMPACT OF IOT IN CONTEXT TO COVID-19

The integrated network for an efficient data transfer and sharing is used by the IoT, as discussed here. It also encourages social workers, patients, and civilians to address each problem and collaboration in conjunction with the service benefactors. The successful tracing of patients and reported cases can therefore be fully guaranteed by using the suggested IoT strategy in the COVID-19 pandemic. Most of the citizens are now conscious of the signs associated with the coronavirus (CoVs). The detection of the cluster can be substantially accomplished by the creation of a knowledge-able community of a linked network. There is also the potential to create certain special device apps in order to help the vulnerable benefit from this [24]. The monitoring unit, such as surgeons, paramedics, caretakers, etc., must adjust the correct documentation and treatment of symptoms, so that the impressive change can be opted out to maximize the overall quarantine.

7.5.1   GLOBAL TECHNOLOGICAL ADVANCEMENTS TO RESOLVE COVID-19 CASES RAPIDLY

Therefore, the Government of India unveiled a mobile app called Arogya Setu, which aims to connect the main possible health facilities with the citizens of India, to resolve and increase awareness of the COVID-19 pandemic [25]. Similarly, for its people, China is releasing its smartphone application called- "Near Touch." This inquiry informs the app holder about the corona-positive person's vicinity. That you will take extra caution when you pass out. At the end of April 2020, the US administration will shortly announce a related form of smartphone application for its residents (Figure 7.4).

Taiwan had the largest number of COVID-19 cases after China. Taiwan, however, rapidly militarized and introduced unique strategies to classify, eliminate, and provide services for the wellbeing of the population on a potential basis. Taiwan has supplied its national health insurance database and merged it with its immigration service, has taken the catalog to instigate big data for analytics, and has created real time alerts based on the travel records and medical symptoms to assist the detection of cases during a medical visit [26]. It also employed this new technology, including QR scanning, related transportation background monitoring, etc., to identify the infected individuals.

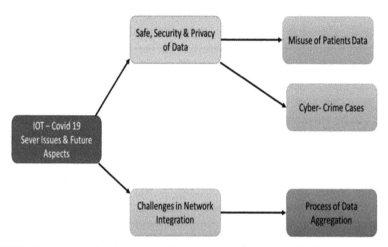

FIGURE 7.4   Summarized view of issues and challenges in implementing IoT for COVID-19.

7.5.2   SIGNIFICANT APPLICATIONS OF IOT FOR COVID-19 PANDEMIC

IoT uses a wide variety of connected devices to build a smart health management system network. It warns and monitors each disease category to optimize the patient's wellbeing [27]. The patient's data and knowledge are digitally registered without human intervention. This knowledge is also important for the related decision-making process. The key IoT implementations for COVID-19 pandemics are discussed in Table 7.1.

TABLE 7.1   Applications of IoT for COVID-19 Pandemic

Sl. No.	Applications	Description
1.	Hospital with Internet access	IoT is a highly developed network of hospital environments to help a pandemic such as COVID-19
2.	Inform the medical professionals involved in any emergency	This interconnected network would allow patients and staff to react quicker and more efficiently as appropriate.

TABLE 7.1 *(Continued)*

Sl. No.	Applications	Description
3.	Transparent therapy with COVID-19	Without partiality or favors, patients may benefit from the benefits.
4.	Online method of treatment	The choice of recovery methods is efficient and tends to manage patients properly.
5.	Advising on telehealth	In fact, the use of well linked teleservices makes the care possible for people in remote areas.
6.	Wireless network for patient recognition COVID-19	Different authentic apps may be built-in smartphones, making a smoother and more efficient recognition procedure.
7.	Intelligent follow-up with untreated patients	The effect of patient tracing finally improved the customers' intelligent handling.
8.	Impact details as the infection is spreading	Successful computer, places, networks etc. knowledge sharing is feasible, and circumstances should be properly handled.
9.	Fast screening of COVID-19	When the first-instance case arrives/founds, the correct decision is made by intelligent wired care systems. At the end, this increases the overall screening process.
10.	Innovative solution defines	The cumulative supervision standard is the ultimate priority. It can be accomplished by effectively applying technologies on the field.
11.	Link all medical equipment and instruments through the internet	IoT linked all medical equipment and instruments thru the internet during COVID-19 care, to provide real-time care information.
12.	Precise Virus Prediction	The use of any predictive approach could also help forecast the situation in the next few years, depending on the data report available. It would also help to plan a safer working environment for the administration, doctors, educators, etc.

7.6 IoT MARKET GROWTH

IoT is supposed to be broadly implemented in a range of industries, beginning with vehicles, and expanding into intelligent residences, intelligent houses, industrial plants and processing plants, health, and safety, security, infrastructure, transport, electricity, distribution, agricultural HEMS, smart meters, etc. IoT products and accessories are available lately with AI headphones, smartphones, intelligent meters, surveillance cameras and more [28].

In addition, IoT is projected to expand much more explosively due to the advent of 5 G networks, which have high and rapid bandwidth, low latency, and multi-connectivity.

On the IoT market, a link between several "thing" (IoT devices and products) and the internet is used to gather information from stand-alone or built-in sensors, then distributed over the Internet to AI data centers for review by transmitting the findings over the internet to other IoT devices and products, smartphones, laptops, etc. The links are made to IoT systems and products [29].

Here are some recent submissions:

- The Internet will send music to listeners from the internet through AI speakers linked to wireless LAN (Wi-Fi TV Wireless LAN). In this case, the listener's history can be stored in a database to be examined by AI to suggest any music that is suited to the listener's taste.
- Smartwatches linked by Bluetooth with the cell phone of a 4 G LTE connect to the internet, helping people with diet and health issues to change their way of living using these details (number of steps and distance walking, cardiogram) from the sensors on the clever watches.

7.6.1   DISSECTING AN OUTBREAK

IoT can be used to detect the cause of an outbreak and detect it. Mobile data from seriously infected patients can do two things-a geographical information system (GIS) overlay on IoT:

1. **Upstream:** Epidemiologists will help tracking patient zero in their quest.
2. **Downstream:** Any patients who could or may not have been in touch with contaminated patients will be detected to decide if they are contaminated.

7.6.2   ENSURING COMPLIANCE WITH QUARANTINE

Chinese safety and monitoring giants have a real-world facility to study and update their inventions, which are not otherwise usable. The evolving high-tech monitoring in China from Robocop glasses to facial recognition distribution machinery ensures that citizens comply with quarantine [30].

7.7   CONCLUSION

Coronavirus 2019 (COVID-19) was a global force that affected people globally and altered the usual aspects of humans significantly. IoT delivers a large and interconnected healthcare network in order to tackle the pandemic COVID-19. Both medical instruments are internet-connected and instantly send a message to the medical personnel in any vital situation. Infected situations of well-connected tele-devices may be handled properly remotely. Both cases are treated intelligently to eventually provide improved medical and health care facilities. The IoT seems to be a perfect way to track the patient sick. Consequently, the technology useful for a pandemic focus predominantly on artificial intelligence (AI) and robots, are obviously listed in this chapter. These include AI, ML, and IoT, robot technologies and invention. The creation of COVID-19 has forced improvement restrictions and various possibilities for the experience of the pandemic have been suggested. This has accelerated innovation in nearly all fields in general. During this time, online learning opportunities can have a long-lasting effect on Internet shopping, e-learning, computerized implementations, telehealth offices and diversion. Artificial intuition, the AI, IoT, advances on areas and roads, enhanced, and computer-generated fact, rocking creativity, the introduction of autonomy, cloud, and diversion technologies will all be incorporated into the center 's growth.

KEYWORDS

- artificial intelligence
- bigdata
- computational technologies
- COVID-19
- internet of healthcare things
- internet of medical things
- internet of things
- medical services internet

REFERENCES

1. Kumar, A., Sharma, K., Singh, H., Naugriya, S. G., Gill, S. S., & Buyya, R., (2020). A drone-based networked system and methods for combating coronavirus disease (COVID-19) pandemic. *Future Generation Computer Systems.*

2. Rhodes, S. D., Mann-Jackson, L., Alonzo, J., Garcia, M., Tanner, A. E., Smart, B. D., & Wilkin, A. M., (2020). A rapid qualitative assessment of the impact of the COVID-19 pandemic on a racially/ethnically diverse sample of gay, bisexual, and other men who have sex with men living with HIV in the US South. *AIDS and Behavior*, 1–10.

3. Chamola, V., Hassija, V., Gupta, V., & Guizani, M., (2020). A comprehensive review of the COVID-19 pandemic and the role of IoT, Drones, AI, blockchain, and 5G in managing its impact. *IEEE Access, 8*, 90225–90265.

4. Natukunda, J., Sunguya, B., Ong, K. I. C., & Jimba, M., (2020). Adapting lessons learned from HIV epidemic control to COVID-19 and future outbreaks in sub-Saharan Africa. *Journal of Global Health Science, 2.*

5. Sovacool, B. K., Del, R. D. F., & Griffiths, S., (2020). Contextualizing the COVID-19 pandemic for a carbon-constrained world: Insights for sustainability transitions, energy justice, and research methodology. *Energy Research & Social Science, 68,* 101701.

6. Stylidou, A., Zervopoulos, A., Alvanou, A. G., Koufoudakis, G., Tsoumanis, G., & Oikonomou, K., (2020). Evaluation of epidemic-based information dissemination in a wireless network testbed. *Technologies, 8*(3), 36.

7. Litman, T., (2020). *Pandemic-Resilient Community Planning.* Victoria Transport Policy Institute.

8. Scott, I. A., & Coiera, E. W., (2020). Can AI help in the fight against COVID-19? *The Medical Journal of Australia*, 1.

9. Thakur, N., Lovinsky-Desir, S., Bime, C., Wisnivesky, J. P., & Celedón, J. C., (2020). The structural and social determinants of the racial/ethnic disparities in the US COVID-19 pandemic: What's our role? *American Journal of Respiratory and Critical Care Medicine (ja).*

10. Bakaeen, F. G., Gillinov, A. M., Roselli, E. E., Chikwe, J., Moon, M. R., Adams, D. H., & Svensson, L. G., (2020). Cardiac surgery and the coronavirus disease 2019 pandemic: What we know, what we do not know, and what we need to do. *The Journal of Thoracic and Cardiovascular Surgery.*

11. Bashshur, R., Doarn, C. R., Frenk, J. M., Kvedar, J. C., & Woolliscroft, J. O., (2020). *Telemedicine and the COVID-19 Pandemic, Lessons for the Future, 26*(5).

12. Elhassan, R., Sharif, F., & Yousif, T. I., (2020). *Virtual Clinics in the Covid-19 Pandemic, 113*(7), 127.

13. Kricka, L. J., Polevikov, S., Park, J. Y., Fortina, P., Bernardini, S., Satchkov, D., & Grishkov, M., (2020). Artificial intelligence-powered search tools and resources in the fight against COVID-19. *EJIFCC, 31*(2), 106.

14. Maeder, A., Bidargaddi, N., & Williams, P., (2020). Contextualizing digital health contributions to fighting the COVID-19 pandemic. *Journal of the International Society for Telemedicine and eHealth, 8,* e3-1.

15. Shu, M., & Li, J., (2020). Health digital technology in COVID-19 pandemic: Experience from China. *BMJ Innovations*.
16. Bhatia, V., Mandal, P. P., Satyanarayana, S., Aditama, T. Y., & Sharma, M., (2020). Mitigating the impact of the COVID-19 pandemic on progress towards ending tuberculosis in the WHO south-east Asia region. *WHO South-East Asia Journal of Public Health, 9*(2), 95–99.
17. Dosaj, A., Thiyagarajan, D., Ter, H. C., Cheng, J., George, J., Wheatley, C., & Ramanathan, A., (2020). Rapid implementation of telehealth services during the COVID-19 pandemic. *Telemedicine and e-Health*.
18. Baker, C. M., Campbell, P. T., Chades, I., Dean, A. J., Hester, S. M., Holden, M. H., & Possingham, H. P., (2020). *From Climate Change to Pandemics: Decision Science can Help Scientists Have Impact.* arXiv preprint arXiv:2007.13261.
19. Huri, E., & Hamid, R., (2020). Technology-based management of neurourology patients in the COVID-19 pandemic: Is this the future? A report from the International Continence Society (ICS) Institute. *Neurourology and Urodynamics*.
20. Shoptaw, S., Goodman-Meza, D., & Landovitz, R. J., (2020). Collective call to action for HIV/AIDS community-based collaborative science in the era of COVID-19. *AIDS and Behavior*, 1.
21. Simons-Rudolph, A., Lilliott-González, L., Fisher, D. A., & Ringwalt, C. L., (2020). *The Effects of the COVID-19 Pandemic on the City Pilots Served by AB InBev Foundation's Global Smart Drinking Goals Initiative, 10*, 1–12.
22. Lee, D., & Lee, J., (2020). Testing on the move South Korea's rapid response to the COVID-19 pandemic. *Transportation Research Interdisciplinary Perspectives*, 100111.
23. Cheng, P., Xia, G., Pang, P., Wu, B., Jiang, W., Li, Y. T., & Dai, X., (2020). COVID-19 epidemic peer support and crisis intervention via social media. *Community Mental Health Journal*, 1.
24. Bajpai, N., Biberman, J., & Ye, Y., (2020). *ICTs and Public Health in the Context of a Pandemic.* https://doi.org/10.7916/d8-hbbh-e863 (accessed on 14 July 2021).
25. McLoughlin, G. M., McCarthy, J. A., McGuirt, J. T., Singleton, C. R., Dunn, C. G., & Gadhoke, P., (2020). Addressing food insecurity through a health equity lens: A case study of large urban school districts during the COVID-19 pandemic. *Journal of Urban Health*, 1–17.
26. D'cruz, M., & Banerjee, D., (2020). 'An invisible human rights crisis': The marginalization of older adults during the COVID-19 pandemic: An advocacy review. *Psychiatry Research*, 113369.
27. Malik, S., & Naeem, K., (2020). *Impact of COVID-19 Pandemic on Women: Health, Livelihoods & Domestic Violence.* http://hdl.handle.net/11540/11907 (accessed on 14 July 2021).
28. Grundy-Warr, C., & Lin, S., (2020). COVID-19 geopolitics: Silence and erasure in Cambodia and Myanmar in times of pandemic. *Eurasian Geography and Economics*, 1–18.
29. Ros, M., & Neuwirth, L. S., (2020). Increasing global awareness of timely COVID-19 healthcare guidelines through FPV training tutorials: Portable public health crises teaching method. *Nurse Education Today*, 104479.

30. Fefferman, N. H., DeWitte, S., Johnson, S. S., & Lofgren, E. T., (2020). *Leveraging Insight from Centuries of Outbreak Preparedness to Improve Modern Planning Efforts*. arXiv preprint arXiv:2005.09936.

CHAPTER 8

Pre-Detection and Classification of Coronavirus Disease by Artificial Intelligence and Computer Vision

RAJESH V. PATIL, ABHISHEK M. THOTE, and SANDIP T. CHAVAN

School of Mechanical Engineering, Dr. Vishwanath Karad MIT World Peace University, Pune, Maharashtra, India,
E-mails: patilraje@gmail.com (R. V. Patil), abhi.thote8@gmail.com (A. M. Thote), sandip.chavan@mitwpu.edu.in (S. T. Chavan)

ABSTRACT

Unpredictable and speedily scattering the COVID-19 coronavirus moved from its previous attention in China to spread more than 100 countries contaminating over 20 million of people and causing upwards of 15 lakh deaths thus far. Exalted temperature of human body or infection frequently a consistent indicator of numerous infections. When the immune system senses an infection, the main body of inner tear duct temperature increases. Mostly one of the symptoms of COVID-19 is increased human body temperature. For this reason, use of thermal imaging cameras becoming essential measures to control the pandemic and detect possible infections of body temperature in crowded public spaces. However, in facial detection need to sense the temperature as well as to recognize objects of any external heat sources. Furthermore, to overcome the problem of reviewing millions of computed tomography (CT) images per day by recognizing lung ground glass opacities and variations in density, range, and area to help surgeons for type of diagnoses faster by showing and recapitulate symptoms. As the world challenges this kind of epidemic of coronavirus, numerous have praised artificial intelligence as our sagacious tool. Along with, computer vision developed set of systems such as fast discovery of COVID-19 and monitoring of community distancing observes among public. The proposed work presented artificial intelligence and machine intelligence methods as an essential tool to eliminating coronavirus diseases. It delivers novel and consistent standards for medical care facilities. Owing to limitless capabilities of artificial intelligence, increased from its several systems and methods observed directly benefited to discourse infectious spread of SARS-CoV-2 disease globally. Furthermore, machine intelligence-based technologies also played significant part in reply to coronavirus disease epidemic. It benefited to educate about virus, test possible actions, diagnose individuals, analyze community health impressions, and many more. In addition, observed potential for increase competence, usefulness of health investigation and care ecosystem to advance the excellence of patient care.

8.1   INTRODUCTION

The coronavirus (CoVs) disease initiates from a disease group related by simple acute lung patterns and common cold. However, researcher claimed that the CoVs arises due to bats and snakes subsequently circulate and contaminated among peoples. The central east lung pattern is one more contamination alike root as per CoVs disease and at perils position to humanoid health. According to the world health organization (WHO) common sign of CoVs found such as superfluous coughing, high fever, respiratory built issues and many more. With this input, the worldwide CoVs epidemic threatens the survives of people. To avoid this threat, all places from airports to industries along with transportation locked down because the world public suffering every means all over world. However, after this movement the whole world at isolation but after some days all over isolation attempts found inappropriate as disease circulate after a free air and scattering other portions of every nation day by day. The speedy spreading of CoVs after one to another makes virus noxious with present evaluations. Initially, the major indication of increased humanoid physique temperature for CoVs disease.

The heart and brain blood temperature are typically measured as essential of interior humanoid body temperature. However, the humanoid body controls body temperature to retain it continual by swapping heat with atmosphere using diverse controlling schemes including energy, transmission, convection, evaporation, and the ultraviolet (UV) systems showing produced energy to estimate temperature. Generally, the central body temperature uses humanoid body places that are uncovered to the atmosphere as substitutions. As facial casing comprises blood containers near to exterior casing and also its soreness, most infection showing face-mask areas of finest applicant for screening the central body temperature. Telethermograph schemes, also known as thermal imaging schemes are generally apply to measure exterior casing temperature. These schemes include an UV thermal camera and may have a temperature position cause. The thermal imaging schemes and contact less UV thermometers uses diverse methods of infrared (IR) skill to measure temperature. For this reason, the use of thermal imaging cameras becoming essentially applied to minimize epidemic and sense likely contaminations by capturing physique temperature in full community places and in other divisions too. Basically, thermal imaging is a contactless and non-invasive imaging method used for a varied series of biomedical and non-biomedical applications earlier. Mostly, thermal imaging cameras are devices used to detect human body higher temperatures which show signs of infection and another medical disease for additional examination using contact-free approach. It detects warm air release of things in UV spectrum series to sense heat causes and temperature by novel systems with degree of accuracy having precision of +/–0.2°C maintained at constant temperature. However, along with temperature measurement, demanded more parameter to recognize. To fulfill and avoid the spread of CoVs disease, the author introduced artificial intelligence (AI) and computer vision tools. Both techniques contributed fast discovery of COVID-19 and observed well monitoring of community distancing among public. Using one approach this technique detected and monitor forehead temperatures to increase the efficiency and reducing the risk of contagion. It combines RGB and thermal cameras exclusively identify masked faces and measure temperature above 37.5°C (99.5°F) along with to monitor 300 people per minute. The other approaches also described for further references.

Finally, this chapter reviews examine reports about the deterrence and control of epidemics of COVID-19 by AI and machine vision tools. In Section 8.2 introduced allied investigation of CoVs disease detection and prediction by AI and machine vision. In Section 8.3 presented AI advances for discovery of COVID-19 techniques for disease discovery, disease screening and monitoring, investigational treatments, showing CoVs pneumonia, CoVs disease facts and evidence assembly based on internet on things, resource provision to patients, robot for medical isolation patients and finally preventative medicine using AI and telemedicine. At last, in Section 8.4 found AI as a sagacious tool for technological improvement and helping in curving pandemic for less influenced by society.

8.2   ALLIED INVESTIGATION

COVID-19 is a worldwide task and need to talk by all systematic means. Though, an expert system and computer vision system found prominent for finding in early days of CoVs disease. In direction to detection of COVID-19 by AI, numerous researchers contributed such as seven important claims of AI for COVID-19 pandemic recognized and by detecting group of cases, forecasted wherever virus disturb in forthcoming through gathering and analyzing earlier numbers [1]. The AI systems by improved loaded auto-encoder shown the dynamics epidemics communication by calculating actual dimensions besides finish period of CoVs diseases across China [2]. Moreover, the dual cataloging model offers developed performance dimensions in forecasting confirmed cases. By regression analysis and inclination of confirmed cases linked with the variations of every day's climate constraints. At last, found the relative humidity and extreme every day's temperature impacted highest on confirmed cases [3].

Also, the neural network extracted pictorial structures from upper body computerized tomography (CT) images for finding CoVs disease [4]. The AI developed using two methods including an artificial neural network (ANN) on element group optimization and variance progress for arranging weather besides city limits. In addition, city limit, relative moisture found uppermost to forecast established cases of COVID-19 [5]. The clinical features mixtures of COVID-19 forecasted results and grow tool with AI capabilities to imagine patients at risk for additional severe disease on preliminary appearance [6]. Furthermore, the improved algorithms, especially in terms of interpretability and simplification invented for virus cause forecast, epidemic growth predicting, drug investigation and clinical analysis by AI in battling epidemics [7]. The diverse forecast models such as support vector machine, face book's forecaster, holt-winters, and long-short time reminiscence and found the Facebook's prediction model gives lowermost RMSE number for entire nations [8]. The intelligent retrieval model constructed on convolution neural networks (CNNs) to offered an initial and program discovery of COVID-19 patient to stop feast of virus and found competent [9]. Moreover, the AI observed qualified context to sense smartphone sensors signals to forecast the severity of pneumonia and disease [10]. The developed bottomless learning technique extracted graphic structures of CoVs diseases with the aim of medical analysis ahead of infective trial [12]. The summary of machine learning (ML) and AI challenges numerous features of CoVs diseases crisis at diverse forms presented [11].

The three diverse convolutional neural network techniques for CoVs pneumonia patient discovery by chest radiographs and upper body efficiently revealed [13, 14]. The smart recovery style with place care system for further to classify CoVs disease at upper body radiograph and efficiently added analytic system for front medical surgeons [15]. The differentiator separated CoVs diseases from public attained pneumonia and precisely senses CoVs diseases and distinguish it from public attained pneumonia and lung viruses [16]. The quantification of lung opacification restrained on upper body CT method eliminated partiality in early valuation and carry out of lung conclusions in CoVs diseases [17]. Finally, the intelligent retrieval scheme contributed to analyzing possible huge quantity of thoracic CT tests. On the basis of radiology discoveries and earlier review of COVID-19 in details, an intelligent retrieval structure sorted unique CoVs diseases from cold germ linked pneumonia [19, 20].

8.3 ARTIFICIAL INTELLIGENCE (AI) ADVANCES FOR DETECTION OF COVID-19

AI used as nutrition apparatus to battle versus disease-causing epidemic that affected entire globe since beginning of 2020. However, it has been deployed as powerful weapons against COVID-19 virus from transmission, analysis to containment and medicine development. With its capacity to acquire speedily from facts linking to new CoVs, it protects humanoid lives period by ordering genome of SARS-CoV-2, scheming laboratory trials, categorizing computerized axial tomography (CAT) images, faster diagnoses and manufacture original serums. In an epidemic, period plays as a kernel whereas CoVs diseases upsurge rapidly and day by day protected thousands of survives. Basically, classification of genome is the key source for presence identification of virus through lab test design. Regrettably, the novel CoVs found similar like Sars and Mers those susceptible to alteration and rigid to test. With the new CoVs of SARS-CoV-2, numerous scientists classify genome using AI and found whole-genome detected quicker and improved than traditional nucleic-acid method to senses only a portion of a genome. In addition, laboratory tests, CAT lungs pictures of an active mode also sense the symbols of diseases contamination. During widespread, radiotherapists overawed by thousands of pictures inspect each day. Using AI, the system detected CoVs cases by correctness up to 96%. The biotechnology built on messenger ribonucleic acid (mRNA) for which the study of protein folding is important and AI achieved reduce the time mandatory to develop a model vaccine testable on people. The advanced linear fold algorithm forecasts the ribonucleic acid (RNA) construction of a virus to recognize virus attacks cells with lessen time from 55 minutes to 27 seconds and develop a vaccine with advanced steadiness and effectiveness. Using AI-based drug investigate platform of past data offer medicine for present SARS-CoV-2 disease and robotically responding enquiries from the public and counseling people about essential to showing in clinic or homestay for 14 days isolation.

Furthermore, AI-based robots majorly used to sterilize isolated wards, distribute food, medication, and check body temperatures. Along with, autonomous vehicles play a valuable part in providing access to essential supplies for health-care specialists and the public alike those are transporting goods in diseased parts and sterilizing hospitals, also efficiently minimalizing person-to-person transmission and finally lessening the deficiency of medical staff. It also tracks whether citizens follow self-quarantine instructions and those individuals disobeyed the order and left home recognize by the facial recognition system. Moreover, the AI-powered pneumonia screening and symptoms detection system speedily sense and recognize pneumonic symptoms while providing numerical valuation for analysis information such as amount, size, and quantity of pneumonic symptoms. The disease reveals less than one minute with detection accuracy of 92% and recall rate of 97% on exam data groups. In addition, many susceptible peoples are in areas with inadequate medical resources subsequently AI develop online care access. The AI-based consultation platform provided important facilities to persons who need to check health experts about COVID-19 without pushing themselves or others at peril of receiving diseased. By AI-based robocall platform, made over 3 million automatic 1500 phone calls in one second demanding people willingly to offer their current travel data, nearby contacts, and present health situations. Basically, health commands chased the spread of CoVs across exact areas and help medical alertness from a local perspective using this platform. The AI and big data used to better recognize key trials and places associated to CoVs, identify its origin and rate of spread.

The AI-powered data analytics technique pulls understandings from online behavior such as online search enquiries and social media discussions to recognize signals from exact people to deliver insight as to measure the CoVs. In addition, AI-powered mapping systems identified the flow of travel across high-risk areas. It assisted epidemiologists build an estimated portrait of people's relocation with around carrying CoVs. It also provided users and healthcare specialists with valued real-time understanding into CoVs's spread to accelerate local preparation and response efforts. The AI system recognizes CoVs with an accuracy to be 96% and develops around 300 to 400 desirable images to identify a CoVs in 20 to 30 seconds, whereas the same process typically takes 10 to 15 minutes by skilled doctors. Figure 8.1 shows probable solutions by AI for detection of COVID-19, and Table 8.1 presented country-wise development of AI tool for detection of COVID-19.

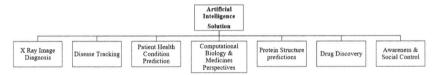

FIGURE 8.1   Probable solutions by artificial intelligence for detection of COVID-19.

8.3.1   ARTIFICIAL INTELLIGENCE (AI) FOR DISCOVERY OF ALLEGED CORONAVIRUS DISEASES CASES

An epidemic generates inimitable hurdles to distribution of medical care and necessity to confronted an inadequate quantity of medical care workforces. In spite of said, AI help to address these difficulties such as, a smartphone application developed to gathers ciphers, indications, earlier places of patient, transportable past, and restructured zones of epidemic, after procedures and sieves this evidence by state algorithms to examined alleged cases through surgeons. Furthermore, AI plays the role of isolated nursing for home isolated patient besides their relatives over smartphones. A program ring provides a cautionary communication when patient or household fellow will breakdowns isolation. Figure 8.2 shows AI discovery for alleged CoVs diseases cases.

TABLE 8.1 Country Wise Development of Artificial Intelligence Tool for Detection of COVID-19

Country	Developed Artificial Intelligence Applications	Functionality
Argentina	COVID-19 Ministerio de Salud	Self-diagnostic
	Coronavirus Australia	Quarantine Enforcement
	CORONAlert	Alerting
	COVIDSafe, StoppCorona	Contact tracing
Bahrain	BeAware Bahrain	Quarantine Enforcement
Brazil	Coronavírus-SUS	Information
	The Spread Project	Contact tracing
Bulgaria	ViruSafe	Contact tracing
Canada	Canada COVID-19	Self-diagnostic
	COVID Shield, Covi	Contact tracing
Chile	CoronApp	Self-diagnostic
China	Alipay Health Code	Contact tracing
Colombia	CoronApp-Colombia	Medical reporting
Czech Republic	eRouška	Contact tracing
Denmark	Smittestopp	Contact tracing
Finland	Ketju	Contact tracing
France	StopCovid, uTakeCare, Alertanoo	Contact tracing
Georgia	STOPCovid	Contact tracing
Germany	Coronika, Our Health in Our Hands (OHIOH), Ito	Contact tracing
Ghana	GH COVID-19 Tracker App	NA
Greece	DOCANDU COVID Checker	Self-diagnostic
Hong Kong	Stay Home Safe	Quarantine enforcement
Hungary	VirusRadar	Contact tracing
Iceland	Rakning C-19	Contact tracing
India	Test Yourself Goa, Test Yourself Puducherry, Trackcovid-19.org	Self-diagnostic
	Corona Watch, Mahakavach, COVA Punjab, Aarogya Setu, COVID-19 Quarantine Monitor	Contact tracing
	Quarantine Watch	Quarantine enforcement
	COVID-19 Feedback	Medical reporting

TABLE 8.1 *(Continued)*

Country	Developed Artificial Intelligence Applications	Functionality
Indonesia	PeduliLindungi	Contact tracing
Israel	Hamagen	Contact tracing
Italy	allertaLOM	Medical reporting
	diAry "Digital Arianna," Immuni, Rintraccia dei contatti, SM-COVID-19, CovidApp-COVID Community Alert,	Contact tracing
Jordan	AMAN	Contact Tracing
Kuwait	Shlonik	Self-diagnostic
Latvia	Apturi COVID	Contact Tracing
Malaysia	Gerak Malaysia, MyTrace	Contact Tracing
	MySejahtera	Information
Morocco	Wiqaytna	Contact tracing
Mexico	Plan Jalisco COVID-19	Contact tracing
	COVID-19MX	Self-diagnostic
Netherlands	PrivateTracer	Contact Tracing
North Macedonia	StopKorona	Contact tracing
Norway	Smittestopp	Contact tracing
Poland	Kwarantanna domowa	Quarantine enforcement
	ProteGO	Contact tracing
Qatar	COVI	Information
Republic of Angola	COVID-19 AO	Quarantine enforcement
Russia	Social Monitoring	Contact tracing
Saudi Arabia	Tawakkalna (COVID-19 KSA)	Quarantine enforcement
Singapore	TraceTogether, SafeEntry	Contact tracing
South Africa	Covi-ID	Contact tracing
South Korea	Self-Isolator Safety Protection, Self-Quarantine App	Quarantine enforcement
	Mobile self-diagnosis	Self-diagnostic
Spain	STOP COVID-19 CAT	Information
	COVID-19.eus	Contact tracing
	CoronaMadrid	Medical reporting
Sri Lanka	Self-Shield	Quarantine enforcement

TABLE 8.1 *(Continued)*

Country	Developed Artificial Intelligence Applications	Functionality
Switzerland	SwissCovid	Contact tracing
Thailand	MorChana	Contact tracing
Turkey	Korona Önlem	Self-diagnostic
United Arab Emirates	Tawakkalna (COVID-19 KSA)	Quarantine Enforcement
United Kingdom	COVID Symptom Study	Medical reporting
	NHS App	Contact tracing
United States	Coalition App, CovidSafe, Private Kit: Safe Paths, COVID Watch, NOVID	Contact tracing
	COVID-19 Apple App	Information
	How We Feel	Self-diagnostic
	coEpi	Medical reporting
Uruguay	Coronavirus UY	Self-diagnostic
Vietnam	COVID-19	Self-diagnostic

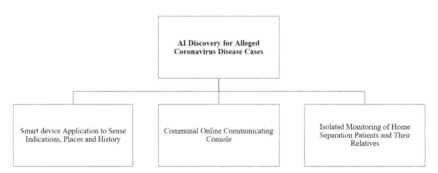

FIGURE 8.2   Artificial intelligence discovery for alleged coronavirus diseases cases.

8.3.2 ARTIFICIAL INTELLIGENCE (AI) FOR CORONAVIRUS DISEASES SCREENING

The contactless schemes use AI for quantity indications and signs of CoVs diseases significantly for huge scale transmission in purposely less time. Fast showing by these systems allows to remote discovery of alleged cases along with achieved by knowingly less people, decrease work on zones of airfields, superstores, and more augment communal disaffection [20]. In contactless schemes for huge scale showing camera-based motion discovery software and IR thermal imaging technology to sense and examine irregular respiratory forms. The IR radiation thermal photographic camera allows actual time imagining of momentary along with continuous variations in energy glowing from persons which permits external temperature approximation [21]. With AI discovery systems, alleged persons with COVID-19 routinely recognized and followed by IR thermal cameras with negligible requirement of humanoid monitoring. The breathing form found added investigative symbol of COVID-19 patients [22].

The form of breathing in COVID-19 is fairly diverse from communal cold and quick breathing strongly specify contamination with CoVs diseases. Using a camera to capture contactless persons breathing dimensions and analyzing dimensions by AI discovery systems found talented technique for added authorization of COVID-19 cases [23, 24]. Figure 8.3 shows AI tools for COVID-19 screening.

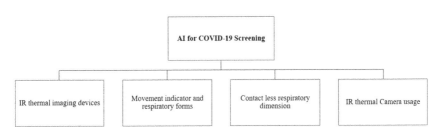

FIGURE 8.3   Artificial intelligence tools for COVID-19 screening.

8.3.3   ARTIFICIAL INTELLIGENCE (AI) FOR CORONAVIRUS DISEASES MONITORING

The amounts of information from patient role with CoVs diseases are physically composed in hospitals with or without separate monitoring equipment and collection procedures of patient display information diverse in many hospital regions. Almost all hospitals information composed physically and noted in worksheets, which wasted once patients get settled. To overcome this problem, the AI technique gathers display data from patient role using private digital aides and alike apparatus; the patients information deposited in electronic health histories where they simply communal and quickly moved when desirable. With diverse technologies and AI, the gathering examination and clarification of patients display information completely or incompletely automated which reduces contagion danger and loads forced on health work to continually collect, supply, examine, and understand this information. Numerous brainy and skilled systems established for medicinal nursing by radio device systems. The approach initiates effective once signals are unremitting, though, pertinence of skilled schemes to inadequate information up till now recognized.

The thoracic calculated imaging (CI) difference careful to be an actual technique to sense, count, and display indications of patients with CoVs diseases. Intelligent retrieval systems established to examination, clarification, and following huge facts of thoracic CI inspections [25]. Emerging video tomography founded high rapidity medicinal nursing scheme practices gesture chasing monitoring systems for affected patient with CoVs diseases offer amount of accurate data concerning dynamic indications along with position, state severity, current comorbidities, and patient release. Average constraints then removed to examine the level of injury from COVID-19. Motion tracking monitoring examined several studies with talented outcomes. It has been stated that patients with COVID-19 settled from the isolation area after fulfilling following criteria: least period of three days, breathing indications determined, developments of radiological lung infiltrates indications at minimum two successive negative CoVs diseases nucleic mordant examinations with sample intermissions of minimum 24 hours period [24].

AI founded automated examination tackles of following release norms for patients with CoVs diseases established to regulate and distinguish treated patients from those who still essential to quarantine. Furthermore, AI computerized CI picture examination tools aimed at following quantification and discovery of COVID-19 assist in distinguishing patients with CoVs diseases persons those have not disease. Additionally, ANNs proficient to conclude qualitative features on strength as well as network implications, which connected giving to patient disorder. An investigative study presented AI dependably professionally donate to precise discovery and following of development or determination of COVID-19. Furthermore, speedy expansions in AI-based automatic tools for CI image examination achieve a high degree of accuracy in discovery of COVID-19 positive patients in adding to detailed valuation of the strictness of the illness [26, 27]. Figure 8.4 shows AI tools for COVID-19 monitoring.

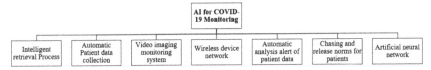

FIGURE 8.4 Artificial intelligence tools for COVID-19 monitoring.

8.3.4 ARTIFICIAL INTELLIGENCE (AI) FOR EXAMINATION OF INVESTIGATIONAL TREATMENTS FOR COVID-19

Computer-assisted medicine design is effective for quickly classifying medicine repurposing applicants. A crucial function of computer-aided drug repurposing giving new infections such as CoVs disease by identifying drugs on their formed to deal with other illnesses. Medicine repurposing understood by showing methodical drug to drug contact examines and reviewing drug-target connections, which can be achieved using AI systems. AI recognized by transformative effect on medicine progress. In accord with current statement machine intelligence and big data's positive effects on health care organizations and optimistic results in medicinal market. Business specialists forecasted as emerging medicines through AI systems deliver meaningfully better responses. An additional division of AI is natural language processing (NLP) useful for CoVs disease medicine expansion. These methods are helpful for takeout sense from machine intelligence scripts and penetrating outside biomedical composition through medicine detection.

Additionally, the AI technique understands details of images used to recognize cell structure to recognize new actions for CoVs disease [28]. The usage of AI is predominant medicine finding to numerous medicinal companies insist businesses with AI firms. Though, data of adequate excellence to trained schemes for CoVs disease AI execution. The data approachability is an added task need to skilled schemes by supervised knowledge desires significant quantities of information on COVID-19 to precisely achieve multifaceted tasks. At present-day, although industry ideals for information recognized for many practices, they presently may not put-on event of COVID-19. Furthermore, a considerable quantity of energy is mandatory to mix information on COVID-19 into company schemes to usage for AI. Figure 8.5 shows AI for investigational examination treatments for COVID-19.

FIGURE 8.5   Artificial intelligence for investigational examination treatments for COVID-19.

8.3.5   ARTIFICIAL INTELLIGENCE (AI) FOR SHOWING CORONAVIRUS PNEUMONIA

CoVs disease scattering quickly at globally in the interim problematic to identify illness. These boundaries partly qualified to inadequate quantity of doctors proficient for understanding information by recognize approaches associated to great increase in quantity of cases [29]. The radiologists differentiate CoVs disease pneumonia from additional forms of pneumonia in upper body CI pictures. This may be due to reduced viral load in the exam specimen or to workshop error. On additional pointer, sensitivity of upper body CI recognized about 58% to 99% for CoVs disease discovery on early demonstration and aid correct untrue adverse outcomes found by Reverse transcription-polymerase chain reaction (RT-PCR) examination through initial growth phases of virus. Moreover, chest CI images reveal illness development [28]. Linked to non-coronavirus disease pneumonia, CI pictures of CoVs disease pneumonia additionally display to vascular deepening, acceptable complicated opaqueness, ground crystal opaqueness, outlying delivery and opposite corona sign in adding two-sided outlying participation of numerous parts. The CI indications recover slowly about fourteen days afterward start signs. On the other hand, CI pictures of CoVs disease pneumonia are improbable to equally bordering display and central delivery [30]. Therefore, forthcoming guidelines in chest CI include emerging an AI classifier additionally supplement presentation of radiotherapist joint with medical data. Automatic examination of pneumonia investigation by intelligent retrieval is significant for inspection of numerous diverse reasons. The most significant motive is that chest radiograms of CoVs disease patients studied by skilled authorities to generate huge quantities of effort for experts. Due to the problems, interpretation chest imageries of CoVs disease patients by humanoid unaccompanied tremendously stimulating [30].

Though arenas of machine intelligence and pneumonia study separately advanced, partial work presently accessible concerning claim of machine intelligence for analysis of pneumonia particularly for CoVs disease patients. Investigation on intelligent retrieval systems newly revealed quick development. In general, pneumonia analysis for medicinal fields is known field. The grouping of two-zone is original and talented. It also stated that machine intelligence has the significant benefit of amplified competence to assess belongings of interferences when perusing sub populaces in scientific pneumonia study [38]. Another important part of machine intelligence for recognizing CoVs disease contagion is image cataloging. This caused in variety expansion of systems for image cataloging. The analytical models for intelligent retrieval determine complicated structures in huge data sets by backpropagation systems to specify mechanism, to adjust interior limits and its usages to calculate each images coating from illustrations of earlier film. Figure 8.6 shows AI tools for showing COVID-19 pneumonia.

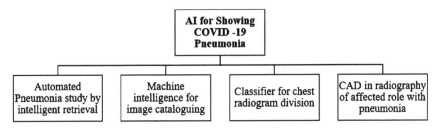

FIGURE 8.6   Artificial intelligence for showing COVID-19 pneumonia.

8.3.6 ARTIFICIAL INTELLIGENCE (AI) AND IOT FOR CORONAVIRUS DISEASE FACTS, EVIDENCE ASSEMBLY

The patients with COVID-19 deliver original sorts of information that are related to attaining medicinal goals. Person chasing instruments, cell phone fitness application and communal television deliver patient's data and permit to display their fitness. Including AI into policy proposals tools and approaches for scheming and analyzing therapies completely available to patients and health worker for supportive into patients survives and health worker follows [31]. This arena proposals are capable to trials for patients with COVID-19 and also tool of advance methodical information gathering to regulate conduct efficiency. In background of internet of smart things, thing is scheme collected of subsystems and each subsystem is a crucial constituent of scheme. Therefore, split CoVs disease medical to investigate policies in fitness hubs or medical places into things and control adequately brainy work on their individual to create behavior for each subsystem and the trial is to comprehend apiece thing distinctly. Therefore, comprehend behavior of things to regulate CoVs disease essentially to know the rise key subparts of apiece thing to acknowledge performance of apiece sub-skill for corrective actions of object an entire to permit and connect with added things through cyberspace. A vital challenge for the progression of numerical health is organized and integration of imperfect data of CoVs disease from diverse causes and changed sorts, such as interoperability resolutions, inadequate accessibility, and existence of present data. All this challenging data is delaying the growth of actual claims.

The diverse data types necessity be combined such as clinical data, info pulls out from bio medicinal works by text mining, and quantity data on medicines interrelate in diverse situations [28]. Moreover, big data found valuable to improve supply usage when making knowledgeable conclusions built on obtainability of data connected to CoVs disease cases. However, AI, internet of things (IoT) and machine intelligence donate meaningfully to process [32]. Big data examination achieved through AI to understood as computers exercise to impersonator thoughtful forms to simulate uniform humanoid behaviors. The attained outcomes accurateness will rise with increasing obtainability of giving out control and information needed to achieve machine intelligence via actual information. Brainy computing systems providing executive for multifaceted difficulties. A contract of achievement attained in participating skilled into brainy systems. Though, skilled systems face problems in obtaining and giving out CoVs disease information. Therefore, to know complicated forms and information added after diverse arenas dynamic to association information removal with brainy calculating schemes and to regulate info collected, which may comprise cluster processes, neural network systems, regression systems and Bayesian processes.

Many other trials arise, counting confidentiality breaks, moral worries, and absence of data safety. Therefore, the capability to segment, examine, and gather info of CoVs disease in actual diverse plans time enhance to struggle of keeping patient confidentiality. Figure 8.7 shows AI and IoT tools for COVID-19 facts, evidence assembly and incorporation.

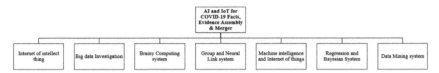

FIGURE 8.7   AI and IoT tools for COVID-19 facts, evidence assembly and incorporation.

8.3.7 ARTIFICIAL INTELLIGENCE (AI) FOR RESOURCE PROVISION TO PATIENTS WITH COVID-19

The entire nations are infected by CoVs disease globally and progressively essential to arrange provision of patient's possessions accordingly [33]. To evaluate traditional approximations of supplies suitable match to CoVs disease however discovered suitable assets away from current volumes of medical hand services. To attain a precise approximation of these resources, approximations should contain humanoid resources, such as doctor's psychotherapists, nurses, and other services counting records of medical hub, emergency sections, ICU, adult, newborn, pediatric table, ventilators, oxygen concentrators, oxygen rolls and plants, fluid oxygen, medicines, individual shielding apparatus, serious medical provisions and beat oximeters. To evaluate all these resources before creating any act plan for resource distribution it significantly demanded. Assumed these facts, arc of diseased persons is compressed over a less period of time and found probable outcomes of CoVs epidemic by lack of medical possessions, mainly ventilators and medical care couches. The affected medicinal employees should also take into thought as these workforces normally be isolated. In adding, even after a serum is established, period will be required to developed, allocate, and manage it; lacks of serum will possibly rise as well. The supply provision problem found essential to comprehend multifaceted information assemblies associated with ordering standards and possessions.

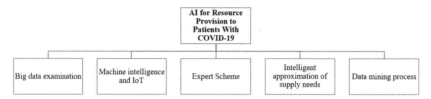

FIGURE 8.8   Artificial intelligence tools for sources provision to coronavirus patients.

Put on big data examination and data mining systems, counting unconfirmed and confirmed machine intelligence improve ordering method supportive to decrease damage in emergency states of uncommon possessions. With the usage of qualified model, machine intelligence minimalizes humanoid energy refining precise supportive arrangement. Moreover, intelligent approximation of source requirements is critical to precise as probable in adding to determining cause of healthier oxygen for patient by seeing entire gross oxygen movement desirable forestalling of individual patient grounded on severity quantity of patients. Evolving a skilled system for practical and reasonable provision of CoVs disease therapeutics and multifaceted possession needs organization of numerous quantities directed by precise data built on IoT technology to offer data about supplies, facts, dimensions, and dangers connected to both resources and pretentious people at global. The understanding of such outlines or skilled systems mostly hinges on distribution and gathering information significantly helped by IoT. Figure 8.8 shows AI tools for sources provision to CoVs patients.

8.3.8   ARTIFICIAL INTELLIGENCE (AI) FOR COVID-19 FORECAST

8.3.8.1   COMMUNICATION FREQUENCY

In the early phases of CoVs disease, epidemic found serious to examine and comprehend the dynamics spread of disease. The variations in approximations of communication over time offer understandings of epidemiologic states to establish efficiency of regulate actions [34]. It examines offer forthcoming forecasts development, contribute danger valuation for nations and direct policy of substitute interferences [35]. These investigates current numerous tasks, exclusively in actual time. Furthermore, with

CoVs disease, presence of signs deferred due to period of development and authorization of cases late in agreement with discovery and checking size. AI methods justification of postponements, indecision, and clearly including delays that outcome from normal history of virus reporting procedures. In adding, separate information sources are inadequate, unfair, or seizure dynamics features of epidemic. Still, indication combination methods suitable with numerous information bases allow healthy approximations of communication dynamics collected raucous data [35].

Numerous issues affect communication dynamics of CoVs disease through nation, such as outer people dimension from positive residence to pretentious capitals, topographical sites, interferences, communal, and financial events, medical care services and ecological collection. The gathering procedure of progressive dynamics deliver numerous visions forms of CoVs disease circulation. In accumulation, improved auto encrypt develop to forecast materialistic quantity of established cases of CoVs disease. By imagining original quantities of epidemic, improved auto encrypt develop with recognized constructions and constraints to forecast sizes of upcoming epidemics and simulate the effect of interferences on the harshness and dimensions of epidemics [34]. Information determined AI approaches proposal real-time prediction methods for approximating and chasing harshness of epidemics, evaluating their path, forecasting their distance and supportive result making by health care employees and administrations.

8.3.8.2   TRANSFORMATIONS

The modeling of chronological data significant dynamically. Earlier research offered ways to insert organic orders into minor dimensional route spaces. Moreover, the harshness and humanity rates valuation of COVID-19 sternness by medical demonstration not at all encounter crucial medical requirements. Therefore, presenting an intelligent retrieval prototypical by number of medical structures to forecast harshness and humanity charges of CoVs disease of important cost. Intelligent retrieval built on calculable CI dimensions of grazes and medical structures of early admission support in forecasting CoVs disease harshness, allow doctors to emphasize patients, plan action procedures and follow-up assessments in advance. However, semantic network presented possible explanation to difficulties faced in program body part division [34]. An earlier study, a novel model estimates to forecast of established CoVs disease. It stated that ungenerous models contain five structures are a perfect amount for forecasting CoVs disease harshness. This communal reversion technique with huge dimensional information protracted approximately to use in logistic regression reproductions for result prediction and existence examination. This method found larger to predictable methods selecting forecasters and permit investigators to trust nominated specific structures into individual signs [22, 37].

Additional study confirmed machine intelligence systems are greater than old-style arithmetic model methods for forecasting humanity in patients with pneumonia. Though, it found that none of the evaluated trials showed overall exact forecasts of patient humanity, and all trials exposed wide differences in presentation on events used. In the current study, investigators recommended a process to forestall humanity rate of CoVs disease patients by precision stretched 90%. Furthermore, machine intelligence technique practice to start an analytical prototypical for primary acknowledgment of censoriously unwell patients built on medical and epidemiologic information found since CoVs disease patient. The employed device for this machine intelligence prototypical grounded on measurable categorization of clinical topographies conferring to criticalness and finally the exposed topographies arranged to obtain understandable medical direction.

8.3.9 ARTIFICIAL INTELLIGENCE (AI) FOR CORONAVIRUS DISEASE MODELING AND REPRODUCTION

Scientific model of diseases and contaminations support to streamline procedure of thoughtful active disease. Numerous researchers used normal differential calculations in biomedicine and hygiene to prototypical and made-up diverse scenarios. Diseases supposed to among many and different organic schemes. Though, contempt their variety, several common trials, procedures found in all viruses, acting as virus-related repetition series, essential for creative contagion. Disturbance of one or more of stages harm or stop spread of disease. Similarly, weakening the infectious particle earlier outbreak cloud an actual means to stop spread. Since beneficial lookout, method of modeling and simulating the molecular side by side active of SARS-CoV-2 in part on each phase in adding to particles themselves significant and managing of contagion through disease. This information compulsory to report developing medicine hardy viral straining and upcoming epidemics of new infective classes alike to COVID-19. In concept, evolving therapeutics that mark sole or numerous stages in the viral repetition series or serious procedures that have restricted measurements for feasible alteration decrease opening for SARS-COV-2 advance fight to managed medicines. Similarly, if simulations aid considerate active and physical basis for CoVs disease medicine battle, antiviral drug improved for alterations.

Many areas of environmental science find an all-inclusive considerate of disease processes, and multidisciplinary methods are mandatory. Reinforced by operational biology developments, AI, and computational systems have arisen as controlling tackles that accompaniment investigational methods with usage of scientific modeling and imitations. In numerous cases, AI, and computational methods assistance link data gaps among trials through writing in diverse sequential and three-dimensional areas in addition to their significant projecting controls. Figure 8.9 shows AI for CoVs disease modeling and reproduction.

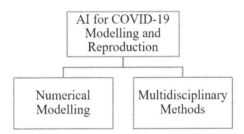

FIGURE 8.9   Artificial intelligence for coronavirus disease modeling and reproduction.

8.3.10   ARTIFICIAL INTELLIGENCE (AI) ROBOT FOR MEDICAL ISOLATION PATIENTS HAVING CORONAVIRUS DISEASE

COVID-19 is a highly contagious virus that attitudes a real danger to health care workforces. Communication of this virus to health care workforces is extremely perspective, particularly during epidemics when hospitals are loaded with diseased patients. AI proposal harmless and effective solutions, such as machines that health care specialists' function while teleconference with patient role. Teleoperated machines can complete common treatment responsibilities in dangerous areas, such as transporting foods or medications, gathering samples, and moving unused, with high precision and proficiency. An obvious benefit of these machines is a sole machinist can regulator numerous machines while quickly converting between isolation areas. Other compensations comprise the skill to connect with patients via a computer-generated telepresence structure twenty-four hours each day. Figure 8.10 shows AI robotics for medical isolation of COVID-19 patients.

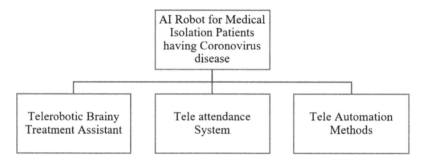

FIGURE 8.10   Artificial intelligence robotics for medical isolation of COVID-19 patients.

8.3.11   TOWARD PREVENTATIVE MEDICINE USING AI AND TELEMEDICINE

Numerous works precisely established the meaning of using tele drug in community health crises and calamities. Tele drug agendas take a period to grow though fitness schemes that have previously established tele drug novelties influence and adjust quickly to manage CoVs disease outbreaks [38]. Onward emphasize measured vital policy of medical care flow controller hinges on mostly previous patient's categorization to their entrance at the emergency section. The customer tele drug is a technique that allows actual transmission of patients. The transmission shields, medical care workforces and public contact also equally patient positioned and helpful for person separation. Tele drug permits patients and doctors to connect by digital devices at any time. At current, main wall to large scale tele drug transmission for CoVs disease is test organization. As accessibility of testing places rises, progress, and incorporation of home-grown schemes into tele drug systems is desirable to test suitable patients lessening contact by shelters, in vehicle testing or devoted workplace. To save stage with emerging positions regarding CoVs disease medicinal schemes recollecting automatic logic flows cite acceptable or perilous patients for treatment and similarly allow patients to request audio-visual calls on-demand [39].

It is significant that observes not regularly mention patients to crucial care health centers or make contact danger for medical care workers and overburden these hubs with affected role. Before the epidemic of CoVs disease, numerous crises in sections familiar their earner in-emphasize representations for quick testing and assessment to permit isolated earners to achieve consumption. In emergency states, web conferencing software with endangered open lines from the emphasize area to a worker can be quickly applied. Using a single isolated clinician to protect numerous sites discourse staff tasks; though, this portion is problematic to instrument if software program absences a line up function. To evade revealing staff, telehealth calls can be directed by balancing profitable systems that enable announcement with workers over devoted networks. Though, this scheme does not completely remove contact of medical care workers to patients need certain physical actions. Electronic nursing agendas allow doctors and nurses to remotely display patient's position in numerous hospitals. Using portable combined medicinal programs, patients able from their households with medicinal provision practically. Tele drug offer quick admittance to authorities who are not punctually accessible in individual. Fences to applying these plans are mostly connected to personal credentials, compensation, and professional employment [36]. CoVs disease has elevated concerning staff capacity. Finally, tele drug allow isolated doctors to remotely attain and desire patients, release time for doctors to deliver individual attention.

8.4 DISCUSSION

The presented AI and machine intelligence methods found essential tool to eliminating CoVs diseases. It delivers novel and consistent standards for medical care facilities. Owing to limitless capabilities of AI, increased from its several systems and methods observed directly benefited to discourse infectious spread of SARS-CoV-2 disease globally. Furthermore, machine intelligence-based technologies also played a significant part in reply to CoVs disease epidemic. It benefited to educate about virus, test possible actions, diagnose individuals, analyze community health impressions and many more. In addition, observed potential for increase competence, usefulness of health investigation and care ecosystem to advance the excellence of patient care. At last, AI along with machine intelligence found as front warriors to defense this CoVs pandemic. Overall praised AI as a sagacious tool to described technological improvement helpful in curving pandemic for less influenced by society and adherence to preventive measures with solutions of foster positive health attitudes.

KEYWORDS

- **artificial intelligence**
- **artificial neural network**
- **computer vision**
- **convolution neural network**
- **COVID-19**
- **machine intelligence**
- **reverse transcription-polymerase chain reaction**

REFERENCES

1. Vaishya, R., Javaid, M., Khan, I. H., & Haleem, A., (2020). *Dia. Met. Syn., Clin. Res. Rev., 14*, 337–339. 10.1016/j.dsx.2020.04.012.
2. Hu, Z., Ge, Q., Jin, L., & Xiong, M., [online] (2020). Available from Cornell University https://arxiv.org/abs/2002.07112 (accessed on 22 June 2021).
3. Pirouz, B., Shaffiee, H. S., & Patrizia, P., (2020). *Sust., 12*(6), 1–21. https://doi.org/10.3390/su12062427.
4. Li, L., Qin, L., Xu, Z., Yin, Y., Wang, X., Kong, B., Bai, J., et al., (2020). *Radi., 296*, E65–E71. https://doi.org/10.1148/radiol.2020200905.
5. Sina, S. H., Behrouz, P., Sami, S. H., Behzad, P., Patrizia, P., Kyoung, S. N., Seo, E. C., & Zong, W. G., (2020). *Int. Jou. of Env. Res. and Pub. Hea., 17*(10), 1–21. https://doi.org/10.3390/ijerph17103730.
6. Xiangao, J., Megan, C., Bari, A., Wang, J., Xinyue, J., Jianping, H., Jichan, S., et al., (2020). *Com. Mat. & Cont., 63*(1), 537–551. doi: 10.32604/cmc.2020.010691.
7. Peipeng, Y., Zhihua, X., Jianwei, F., & Jha, S. K., (2020). *Com. Mat. & Cont., 65*(1), 743–760. 10.32604/cmc.2020.011391.
8. Ramazan, Ü., & Ersin, N., (2020). *Com. Mat. & Cont., 64*(3), 1383–1399. doi: 10.32604/cmc.2020.011155.

9. Mohammad, S., & Mehedi, M., (2020). *Com. Mat. & Cont., 64*(3), 1359–1381. doi: 10.32604/cmc.2020.011326.

10. Maghdid, H. S., Ghafoor, K. Z., Sadiq, A. S., Curran, K., & Rabie, K., (2020). Available from Cornell University http://arxiv.org/abs/2003.07434 (accessed on 22 June 2021).

11. Joseph, B., Alexandra, L., Katherine, H. P., Cynthia, S. N. L., & Miguel, L. O., (2020). Available from Cornell University https://arxiv.org/pdf/2003.11336.pdf (accessed on 22 June 2021).

12. Wang, S., Kang, B., & Ma, J., (2020). *A Deep Learning Algorithm Using CT Images to Screen for Corona Virus Disease (COVID-19)* (pp. 1–28). https://doi.org/10.1101/2020.02.14.20023028.

13. Ali, N., Ceren, K., & Ziynet, P., (2020). Available from Cornell University https://arxiv.org/abs/2003.10849 (accessed on 22 June 2021).

14. Wang, L., & Wong, A., (2020). Available from Cornell University. https://arxiv.org/abs/2003.09871 (accessed on 22 June 2021).

15. Xu, X., Jiang, X., & Ma, C., (2020). Available from Cornell University https://arxiv.org/abs/2002.09334 (accessed on 22 June 2021).

16. Li, L., Qin, L., & Zeguo, X., (2021). *Artificial Intelligence Distinguishes COVID-19 from Community Acquired Pneumonia on Chest CT*. 10.1148/radiol.2020200905.

17. Huang, L., Han, R., Ai, T., Yu, P., Kang, H., Tao, Q., & Xia, L., (2020). *Serial Quantitative Chest CT Assessment of COVID-19: Deep-Learning Approach. Radiology: Cardiothoracic Imaging, 2*(2), e200075. doi: 10.1148/ryct.2020200075.

18. Gozes, O., & Frid-Adar, H., (2020). Available from Cornell University https://arxiv.org/abs/2003.05037 (accessed on 22 June 2021).

19. Ng, M. Y., Lee, E. Y. P., & Yang, J., (2020). *Rad. Card. Ima., 2*(1), 20–34. https://doi.org/10.1148/ryct.2020200034.

20. McCall, B., (2020). *Lan. Dig. Hea., 2*(4), 166, 167. 10.1016/S2589-7500(20)30054-6.

21. Lee, I., Wang, C., Lin, M., Kung, C., Lan, K., & Lee, C., (2020). *J. Hosp. Infect., 105*(1), 102, 103 10.1016/j.jhin.2020.02.022.

22. Huang, C., Xu, X., Cai, Y., Ge, Q., Zeng, G., & Li, X., (2020). *J. Med. Int. Res., 22*(5). 10.2196/19087.

23. Paules, C. I., Marston, H. D., & Fauci, A. S., (2020). *Coronavirus Infections-More Than Just the Common Cold, 23*, 707–708. https://doi.org/10.1148/radiol.2020200370.

24. Pan, F., Ye, T., Sun, P., Gui, S., Liang, B., & Li, L., (2020). *Radi., 295*(3), 715–721. https://doi.org/10.1148/radiol.2020200370.

25. Fang, Y., Zhang, H., Xie, J., Lin, M., Ying, L., & Pang, P., (2020). *Sensitivity of Chest CT for COVID-19: Comparison to RT-PCR, Radi.* https://doi.org/10.1148/radiol.2020200432.

26. Srinivasa, R. A., & Vazquez, J. A., (2020). *Infect. Cont. Hosp. Epid.,* 1–4. 10.1017/ice.2020.61.

27. Robson, B., (2020). *Comp. Biol. Med., 119*, 1–19. 10.1016/j.compbiomed.2020.103670.

28. Dai, W., Zhang, H., Yu, J., Xu, H., Chen, H., & Luo, S., (2020). *Can. Assoc. Radi. J., 71*(2), 195–200. 10.1177/0846537120913033.

29. Fagherazzi, G., Goetzinger, C., Rashid, M., Aguayo, G., & Huiart, L., (2020). *J. Med. Int. Res., 22*(6) doi: 10.2196/19284.

30. Wang, W., Tang, J., & Wei, F., (2020). *J. Med. Virol., 92*(4), 441–447. https://doi.org/10.1002/jmv.25689.

31. Shen, C., Chen, A., Luo, C., Zhang, J., Feng, B., & Liao, W., (2020). *J. Med. Int. Res., 22*(5). doi: 10.2196/19421.

32. Komenda, M., Bulhart, V., Karolyi, M., Jarkovský, J., Mužík, J., & Májek, O., (2020). *J. Med. Int. Res., 22*(5). 10.2196/19367.

33. Emanuel, E. J., Persad, G., Upshur, R., Thome, B., Parker, M., & Glickman, A., (2020). *N. Engl. J Med., 382*(21), 2049–2055. 10.1056/NEJMsb2005114.

34. Wu, J., Leung, K., Bushman, M., Kishore, N., Niehues, R., & De Salazar, P. M., (2020). *Nat. Med., 26*(4), 506–510. 10.1038/s41591-020-0822-7.

35. Kucharski, A., Russell, T., Diamond, C., Liu, Y., Edmunds, J., & Funk, S., (2020). *Lanc. Infe. Dis., 20*(5), 553–558. 10.1016/S1473-3099(20)30144-4.

36. Chen, N., Zhou, M., Dong, X., Qu, J., Gong, F., & Han, Y., (2020). *Lan., 395*, 507–513. https://doi.org/10.1016/S0140-6736(20)30211-7.

37. Hong, Z., Li, N., Li, D., Li, J., Li, B., & Xiong, W., (2020). *J. Med. Inte. Res., 22*(5). 10.2196/19577.

38. Hollander, J. E., & Carr, B. G., (2020). *N. Engl. J. Med., 382*(18), 1679–1681. 10.1056/NEJMp2003539.

39. Nguyen, T., Waurn, G., & Campus, P., (2020). Artificial Intelligence in the Battle Against Coronavirus (COVID-19): A Survey and Future Research Directions, arXiv; preprint arXiv:2008.07343. https://doi.org/10.13140/RG.2.2.36491.23846.

CHAPTER 9

Concept Structure of Database Management System (DBMS) Portal for Real-Time Tracking and Controlling the Spread of Coronavirus

ABHISHEK M. THOTE and RAJESH V. PATIL

Assistant Professor, School of Mechanical Engineering, Dr. Vishwanath Karad MIT World Peace University, Pune – 411038, Maharashtra, India, Phone: +91-8446640525, E-mail: abhi.thote8@gmail.com (A. M. Thote)

ABSTRACT

There are different methodologies established to control any infectious disease. Vaccine has proven to be effective for many patients. However, to control the spread of coronavirus in more efficient manner, the database management system (DBMS) was developed in this study. The database was created with the help of Microsoft SQL Server. The proposed system consisted of governing bodies such as team of DBMS, certified hospitals, corona control cell, police department and travel agencies. Tracking the corona patients and quarantine them after catching for any quarantine rule break will definitely restrict the corona spread. This proposed system will be helpful to provide emergency services to patients through DBMS clients such as corona control cell, police department and hospitals with the help of information provided by team of DBMS. The travel agencies will refrain the currently active corona patients to travel outside the city by accessing their details through DBMS portal. With the due permission of Central or State Government authorities, it can be employed in variety of cities. Team of skilled and trained professionals can make it as a system of prime importance. This database needs efficient software platform to form the complete DBMS portal. This system should be first operated in any one city on a trial basis. After successful implementation practice, it can be launched globally to effectively control the spread of corona virus.

9.1 INTRODUCTION

9.1.1 OVERVIEW OF CORONA VIRUS AND ITS SPREAD

Coronavirus is a single-strand virus RNA with club-type spikes emerging from its surface [1]. The most common symptoms of coronavirus infected person are dry cough, tiredness, and fever [2]. While, symptoms like headache, diarrhea, sored mouth and throat, skin rashes, body pain, pink eye (conjunctivitis) are also observed sometimes in the patient. Most of the patients with mild to moderate infection will experience these symptoms. The serious patients experience pain in the chest, speech loss and difficulty in breathing. Generally, old people and persons suffering from diabetes, respiratory, and cardiovascular problems, etc., are more prone to the coronavirus. Once the person is affected by the coronavirus, it generally takes 5–7 days to show the symptoms or up to 14 days in certain cases [2].

Earlier, two types of Coronaviruses were in existence, namely severe acute respiratory syndrome coronavirus (SARS-CoV) and middle east respiratory syndrome coronavirus (MERS-CoV) [3, 4]. However, the currently globally spread Novel Coronavirus is slightly different than earlier Coronaviruses. Hence, The International Committee on Taxonomy of Viruses (ICTV) has given the scientific name to this Novel Coronavirus as severe acute respiratory syndrome coronavirus 2 (SARS-CoV-2) [5]. It causes Coronavirus disease (also called COVID-19). This Novel Coronavirus was emerged from Wuhan City, China, in December 2019 and it was spread quickly in many countries (all over the world) [6]. The spread of this coronavirus is rising exponentially worldwide but mainly in United States (US), Brazil, and India [7] as shown in Figure 9.1. The X-axis shows number of days from first case of patient and Y-axis shows count of patient. Till 24 August 2020, coronavirus has infected 23.4 million people, out of which approximately 8,09,000 are dead all over the world.

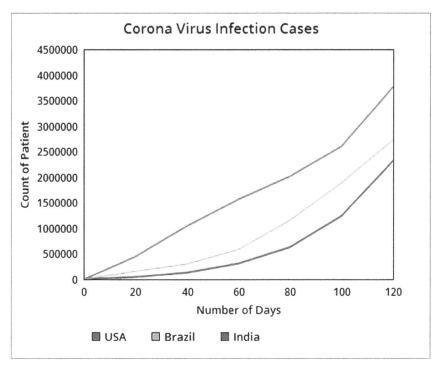

FIGURE 9.1 Exponential rise of coronavirus infection cases in top three most infected countries.

The lifespan of coronavirus is variable on different surfaces. It also remains alive in the air for few hours. Hence, it is also an airborne disease. Based on literature survey [8–11], the lifespan of the coronavirus on different surfaces is represented in Table 9.1.

TABLE 9.1 Lifespan of Corona Virus on Different Surfaces

SL. No.	Surface	Lifespan of Coronavirus
1.	Air	3 hours
2.	Copper	4 hours
3.	Cardboard	24 hours
4.	Stainless steel	2–3 days
5.	Plastic	3 days

No significant reason of emerge of coronavirus is found till date. While, they are commonly found in humans, bats, camels, cats, and cattle. Most probably, it was originated from bats [8]. Coronavirus is contagious (spreads from one sick person to another person). When a sick person sneezes or coughs, the virus spreads through droplets up to the range of 6 feet [8]. Through breathing and swallowing, it infects the healthy person. Some people owing to strong immunity do not show the symptoms, but they can infect the other people with low immunity. Such people are called 'virus carriers.' Coronavirus can stay alive on surfaces of household or outdoor objects. Hence, it also spreads through the eyes, nose, and mouth after touching such virus situated surfaces.

9.1.2   EXISTING CONTROL MEASURES TO PREVENT THE SPREAD OF CORONA VIRUS

To prevent the spread of coronavirus, the following precautionary measures are need to adopt as issued by World Health Organization (WHO) [8, 12]:

- Wear a mask always when going outside home;
- Go outside home only if it is necessary;
- Clean and disinfect the hands regularly with liquid hand wash gel, soap, or alcohol-based sanitizer;
- Maintain a distance of 1–2 meters between two people;
- Cover the face during sneezing or coughing;
- Stay at home if the health is not well;
- Try to avoid physical contact with objects;
- Avoid touching to eyes, nose, and mouth specially when outside the home;
- Do not make group of people, i.e., avoid social gathering;
- Avoid smoking and excessive drinking owing to its harmful impact on lungs;
- Keep only 1 person per 10 square meters of area in the building of workplace;
- Avoid physical business or personal meetings and adopt practice of telephonic or online video conference meetings;
- Do frequent disinfection of commonly used objects, devices floor and wall areas, bathroom, kitchen, light switch surfaces, etc.

- Increase awareness of among employees about coronavirus and it is spread by displaying messages through videos, posters, websites, social networking, and electronic system;
- Carry out regular thermal screening at workplace;
- Undergo 14 days quarantine if in case came in close contact with Corona positive confirmed person;
- Use sodium hypochlorite (bleaching agent) to disinfect the surfaces in work areas with a quantity of 0.1% concentration;
- Use alcohol with minimum 60% concentration to disinfect the surfaces in work areas which may be damaged due to sodium hypochlorite;
- Use medically certified masks, face shields, goggles, disposable gloves, and gowns at workplaces to ensure complete safety.

During outdoor activity or work, the key component to prevent the spread of coronavirus is the use of a face mask. Face mask not only prevents transmission of infectious virus particles from surrounding environment to the person but also from ill person to others [13, 14]. According to the quality of the mask, different masks have variable percentage of effectiveness, i.e., 20% (weakly effective), 50% (moderately effective) and 80–95% (strongly effective) [15].

There are different types of face masks available according to their features and application [16–20]. The common types of face masks are as follows:

1. **Simple Cloth Face Mask:** It is made of cotton or synthetic material. It is generally used by the people while going outdoor for daily work, offices, shops, etc. It is recommended by CDC (Centers for Disease Control and Prevention) to use by the public.
2. **Surgical Face Mask:** It is made of disposable, thin, and nonwoven fabric. It is used by people of medical professionals (doctors, nurses, etc.). It blocks entry of bacteria and germs into the mouth and nose more efficiently. It is recommended for one time use only by FDA (Food and Drug Administration).
3. **N95 and KN95 Respirators or Masks:** N95 and KN95 are similar masks in terms of particles filtration performance. N95 masks are based on the norms of USA standard and KN95 masks are based on the norms of China standard [19]. This is the only difference.

These are specialized masks recommended to use by medical officers, surgeons while operating critical cases of patient. Virus particles are generally in the size range of 0.1–0.3 microns. The efficiency of N95 masks is minimum 95% to prevent these virus particles [20]. During laboratory testing, NaCl (sodium chloride) particles are used to check the filtration efficiency. Nowadays, these masks are also used by the public to prevent the infection owing to the coronavirus.

There are different methodologies established to control any infectious disease. Based on their effectiveness, a hierarchy of controls is specified to take the decision of implementation of a particular method [21–23]. Figure 9.2 shows the hierarchy of controls to be implemented to reduce the spread of any infectious disease. These methodologies vary from most effective to less effective from top to bottom in Figure 9.2. Hence, 'Elimination' is the most effective while PPE (personal protective equipment) is least effective among all other methodologies.

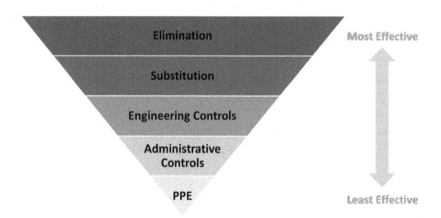

FIGURE 9.2   Hierarchy of controls.

The brief explanation of these methodologies is as follows:

1. **Elimination:** This is the most effective method. It includes completely eliminating the health hazard by avoiding physical contact. For this purpose, people should stay at home unless if

it is really essential and also work from home. Only health care professionals, emergency services, food services should work at respective workplaces obeying all the safety norms. Instead of personal meetings, people should use online meeting applications such as Google Meet, Microsoft Teams, Cisco Webex, WhatsApp, etc.

2. **Substitution:** This is more effective than any other method except elimination. It refers to the replacing the hazardous objects or substances with non-hazardous objects or substances. It also refers to the replacing existing system with the new system. For example, sanitizer shelf in an office should be replaced with a food pedal-operated sanitizing machine or non-contact sensor-based sanitizer dispenser machine.

3. **Engineering Controls:** This is quite an effective method but less effective than elimination and substitution. The engineering control includes several control measures. To control the spread of the coronavirus in offices or homes, EPA (Environmental Protection Agency) registered disinfectants must be used [24]. Commonly used disinfectant is sodium hypochlorite owing to its suitability for residential, office, institute, and healthcare places. Barriers, partitions, protective films should be used for employees working in cabinets of offices. Non-contact trash collection system, sensor operated doors and lighting system, etc., should be incorporated. Highly contacted points like window and table knobs, door handles should be made of copper or at least should be coated with copper tape. The reason is that, coronavirus does not remain alive for more than 4 hours on copper surfaces.

4. **Administrative Controls:** This is a moderately effective and essential method but less effective than the aforementioned three methods. It includes circulation of rules, regulations, or guidelines to be followed by people in their daily routine. Hands should be washed regularly after every work for at least 20 seconds. Hand-sanitizer with more than 60% alcohol should be used in case hand-washing facility is not available, i.e., generally in traveling. Floor marking stickers should be pasted at shops, offices, or any other public gathering places to implement social distancing. Employee health monitoring systems should be used like thermal screening unit, regular health checking camps, etc.

5. **Personal Protective Equipment (PPE):** This is the last but not least method in the hierarchy of controls. This is moderately effective and essential method for personal protection from the surrounding environment. However, this method is less effective than the aforementioned four methods. Employees should be protected with PPE kits. It includes a full body-covering suit with attached cap, eye goggle, face shield, disposable masks, surgical gloves, shoe covers, and disposable bags. It should be mainly used by employees who are frequently interacting with people like doctors, nurses, flight attendant (air host or hostess), cashier in bank, person sitting at counter of office interacting with people in a queue, etc.

Still, there is no vaccine available to cure Corona Virus. Some of the vaccines are under developing stage or at the clinical trial stage. Hence, currently, patients are treated with antibiotics and vitamin tablets only. Immunity of the person is a key factor to fight with Coronavirus disease. Hence, many pharmaceutical companies have developed immunity booster tablets and syrup based on Allopathic, Ayurvedic, and Homeo-pathic medical treatment methodology. Thus, unless effective vaccine launch into the market and this virus vanishes completely from all over the world, control measures must be taken to restrict the community spread of coronavirus. For this purpose, proper, and effective Corona control system must be employed by any country adhering to the rules and regulations of state and national government. This system should be helpful to control the spread of coronavirus in a more efficient manner.

9.1.3 DATABASE MANAGEMENT SYSTEM (DBMS) AND ITS IMPORTANCE

A database is a data storage hub or warehouse where data of persons, products, and people is stored in a structured and sequential manner [25, 26]. Database management system (DBMS) is a software platform which access the database. The users of the DBMS are called clients. The clients can access the DBMS to retrieve the data from the DBMS, but in a controlled manner. DBMS securely provides the data to the clients with certain login and password credentials. The DBMS software system is programmed in such a manner that it provides only specific type of a data based on the type of client. For example, only the design engineer of a particular organization can view or edit the detail design parameters of a product on which he or she is working and can update the database. In another example, the quality engineer can only update the product quality data after inspecting the part or product. For an organization, there are different types of clients such as engineers, workers, suppliers, customers, marketing persons, etc. They access the same database through DBMS but only specific data can be retrieved or updated according to authority granted to a client. In the recent years, as organizations are growing rapidly, they need to store large amount of data. Hence, the use of DBMS is nowadays widely adopted by most of the organizations. The DBMS system is applicable in different fields such as airlines, railways, buses, banks, sales, engineering industries, telecommunication services, educational systems, finance, economics, human resources (HR), etc., [27].

The primary features of DBMS [28] are as follows:

- Quick data sharing from server to the clients and vice versa through intranet or internet;
- Data consistency and uniqueness of data;
- Advanced data security through login credentials like password, face scan, fingerprint scan, one time password (OTP) on mobiles and e-mail addresses;
- Enhanced data display features through better graphical user interfaces;
- Less Maintenance of DBMS software.

The primary components of DBMS (Figure 9.3) are as follows:

- Data of database;
- Software;
- Hardware;
- Process;
- Query language of database.

The clients (users) enter and retrieve the data from the database software using query language. Query language is a simple programming language. The process of data entry and manipulation is defined in the software's program. This DBMS software is installed on the server computer (hardware) where the database is maintained [29].

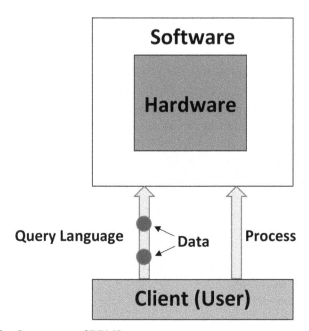

FIGURE 9.3   Components of DBMS.

There are different types of database handling persons [29] classified as follows:

1. **Database Administrator (DBA):** It controls and manages all the data of a database and helps to keep efficient running of database server. DBA grants authority of specific part of the database to clients (users) according to their types.
2. **Software and Maintenance Engineer:** These engineers develop and test the software program (codes) of DBMS which access the database.
3. **Clients (Users):** These are the persons who add, delete, update, or retrieve the data from the database using DBMS software. As mentioned in the first paragraph of this section, there are different types of clients working in several fields.

DBMS organizes the data with a particular logic and structure which is called DBMS model. There are different types of DBMS models such as hierarchical, network, entity-relationship, and relational [30]. Out of these models, relational DBMS (RDBMS) model, also called RDBMS is widely used. In RDBMS, the data is stored in the form of two-dimensional tables with different attributes (properties) stored in columns of the table. One table is linked to another with a common attribute. In the rows of the table, the all the information (attributes) of a particular component or product is stored [30]. Thus, tables are called relations in RDBMS. The different RDBMS software are currently available such as MySQL, Microsoft SQL Server, Oracle database, IBM Db2, Amazon Aurora, PostgreSQL, Amazon relational database service (RDS), IBM Informix, Google Cloud SQL, Maria DB, SQLite, memSQL, etc., [31].

9.1.4   OVERVIEW OF PROPOSED STRUCTURE OF DATABASE MANAGEMENT SYSTEM (DBMS)

The spread of COVID-19 pandemic (coronavirus) is expanding exponentially day-by-day as shown in Figure 9.1. The mortality rate is also rising rapidly. All the safety norms prescribed by WHO are not followed by people. Additionally, the spread is increasing owing to not obeying the rules of social distancing or quarantine by Corona positive patients. Hence, real-time tracking and controlling of Corona patients is a crucial task to break the chain of Corona spread. So, in this chapter, the structure of an efficient DBMS is proposed. Also, it is suggested that the special access to police department, hospitals, Corona Control Cells should be provided with legal permission.

The objectives of proposed structure of DBMS are mentioned as follows:

- To collect the basic information (data) of the tested Corona positive patients by providing database client access to the certified hospitals.
- To collect detailed data of the patient by the team of DBMS.
- To collect the data of the recovered or dead Corona patients from the hospitals and updating the DBMS.
- To provide the data of currently positive Corona patients to the travel agencies to refrain such patients from traveling.

- To give database client access to the Corona Control Cell with information of Corona positive patients and with real time tracking features of a particular person.
- To check by Corona Control Cell if any Corona positive patient is breaking the rule of quarantine and inform to the police department to catch him/her for the purpose of re-quarantine.
- To give database client access to the police department about the necessary information of the Corona patient with real time tracking features of a particular person.

This proposed structure of DBMS for real-time tracking and controlling of Corona patients falls in the methodology of 'engineering controls' in the hierarchy of controls (Figure 9.2). The detailed structure of this proposed DBMS portal is explained in Section 9.2.

9.2   MATERIALS AND METHODS

9.2.1   GOVERNING BODIES

The proposed structure of DBMS needs different governing bodies for smooth and efficient flow of information, i.e., from governing bodies to DBMS and vice-versa. With the approval of concerning state or national government, these bodies can be given respective authorities. This structure of DBMS is proposed for a one particular city for better implementation and control. Similarly, it can be implemented in other cities also. The different governing bodies needed for this DBMS are as follows:

1. **Team of DBMS:** It will consist of Team Manager and assistants, System Administrator and assistants, the DBMS software developing and testing engineers. This team of DBMS can be the part of anyone reputed private software organization. Contract can be given by the Government to this software organization. Else, Government also can form dedicated software organization to develop and manage the DBMS software.

   Software developing engineers should only be focused on updating the software features and its graphical user interface. Software testing engineers should only be focused on corrective maintenance (elimination of any flaws or bugs in the running software) and preventive maintenance (provision of data security, malware check and removal of junk files). DBA and his/her team should be focused on data storage in the DBMS on a daily basis and give client access to the required governing bodies, i.e., users. Manager and his/her team should keep track on each and every activity in the DBMS organization and should suggest better implementation strategies.

2. **Certified Hospitals:** Government and private hospitals treating Corona patients are generally certified by the city municipal corporation. These hospitals keep the record of Corona patients treated in their hospitals. So, these hospitals can form the one of the governing bodies, i.e., client (user) for the proposed DBMS. DBMS software can be installed on the computers of these hospitals. The team of DBMS will verify certain doctors and nurses in the hospitals. Then, these doctors and nurses will be given client access to enter the data of the under treatment and treated corona patients on a daily basis. This is the only source of getting authenticate information of Corona positive patients.

Additionally, these hospitals will update the status of treated Corona patients, i.e., under treatment, recovered or dead due to Corona. This information will also be updated in the DBMS by the hospitals. This will help to know total infected cases, currently active cases, recovered cases and dead cases in each and every certified hospital.

3. **Corona Control Cell:** To keep watch on each and every Corona active patient, there is a need to form the governing body called 'Corona Control Cell.' The DBMS team should be only focused on collecting all the necessary information of Corona positive patients and keeping the DBMS updated. Nowadays, Corona positive patients are increasing drastically. There may not be enough number of bed system available in the hospitals. Most of the patients are mildly affected. So, hospitals mainly focus on treating severe cases of patients and others also if beds are available. Thus, the mildly affected patients who could not be admitted to hospitals are instructed for home quarantine with necessary medical kit (medicines, oxygen cylinder, etc.). So, doctor or nurse cannot be present all the time to monitor the activity of the patient.

In this case, the DBMS system should separately collect the data of home quarantined as well as hospital quarantined Corona positive patients. This information should be forwarded to Corona Control Cell. Then, Corona Control Cell should track should track these patients through their mobile telecommunication service providers. For this purpose, they should perform rigorous and continuous task of tracing the location of patients one by one. So, according to the number of Corona patients, an appropriate number of staff should be employed in Corona Control Cell. City municipal corporation should employ qualified and dedicated people in Corona Control Cell. If any patient is leaving the area of his home or place of quarantine, Corona Control Cell should immediately inform to police. Then, role of the police department is explained in next section.

4. **Police Department:** It should be provided with client access to retrieve the information of currently active Corona patients. Additionally, they should be able to track the location of mobile phones of patients. However, Corona Control Cell needs to identify first if any currently Corona positive patient is leaving the home or place of quarantine as they are continuously engaged in this activity. As soon as the police department receives information of breaking the rule of quarantine by a particular Corona patient, they should be able to immediately access the data of the respective person through their client login credentials. Thus, they should use mobile tracing feature only for patients breaking the rules of quarantine.

   For this purpose, only specific and verified police authorities should be given client access, or the head of city police can form the special cell of police persons which will access the information of the patient. Finally, police should catch the quarantine rule braking Corona patients and should take the disciplinary actions against the patient as per rules and legislations of a particular country. However, the first task required to do is that police should bring them back to their home or hospital with a warning message. This will surely help to break the chain of Corona infection in the society.

5. **Travel Agencies:** There are different travel agencies such as private car/van, bus, train, and flight, etc. The operators and

ticket booking staff of these travel agencies should be clients (users) of the database. Some travelers book online tickets. So, these travel agencies should provide basic information of travelers entering into their cities and should update the DBMS on a daily basis. This information will be helpful in future. If any new corona patient is detected in city, team of DBMS can easily check whether he/she has travel history or not. So, team of DBMS can get the information of other travelers traveling along with him/her. So, a team of DBMS can suggest those travelers to carry out their Corona test and take the feedback of the test report. This will be helpful to know whether other travelers are Corona positive or not. If report of any other traveler is found to be Corona positive, then it will be informed to Corona Control Cell. So, Corona Control Cell will instruct the other traveler to admit in hospital or if no beds are available in hospitals, he/she will be instructed for self-isolation at home. Corona Control Cell will keep watch on the movements of these patients. In this way, Corona Control Cell will track each and every patient, whether identified by hospitals as well as identified by a team of DBMS with the help of travel agencies.

The travel agencies should also be provided with the information of updated list of currently Corona positive patient. Additionally, travel agencies should also provide the travelers leaving this city. So, if any Corona positive patient in the city has booked the ticket of travel, travel agencies will get the alert message on their client DBMS portal. So, travel agencies should cancel such tickets and refund the money. Additionally, a team of DBMS will take feedback of this alert message in the form of cancellation proof to verify whether they have canceled such ticket or not.

9.2.2   HIERARCHY OF INFORMATION FLOW

For any flow of information, hierarchy must be defined. It segregates the authorities from lower level to higher level to pass the information. It also helps to understand the importance of any authority for the pass of information and also to control the entire system. The higher-level authorities have more responsibilities for successful accomplishment of work and hence, they have more significance.

For the proposed structure of DBMS, the hierarchy of authorities with respect to governing bodies is illustrated in Figure 9.4.

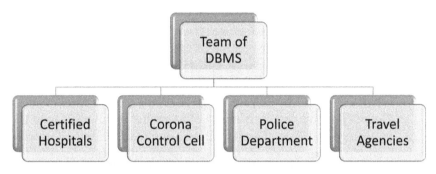

FIGURE 9.4   Hierarchy of information flow.

9.2.3   SCHEMATICS OF INFORMATION FLOW

For better understanding of the flow of information, flow diagram is essential. Figure 9.5 shows schematics of information flow. The detailed pointwise explanation of workflow is stated below:

- Certified hospitals will perform Corona Test of patient admitted to their hospital. If Corona test of a person is found to be negative, then hospitals will enter data of such patients in the DBMS portal under the category of general patient information.
- If Corona test is found to be positive, then hospital or clinic staff will enter data of such patients in DBMS portal under the category of corona patient information. Additionally, DBMS portal will provide the information of positive patients to their clients such as Corona Control Cell, police department and travel agencies.
- Travel agencies will allow ticket booking only to non-corona patients.
- Hospitals or clinics will provide the status of the patient, i.e., under treatment, recovered or dead in DBMS portal on a daily basis.
- If the status of the person is dead, then the DBMS portal will close the further entry of dead person's data by any of the clients of

DBMS. This information will also be made available on DBMS portal.

- If the status of the person is recovered, then the DBMS portal will save the data of these persons separately. However, further entry of these person's data will be allowed in future by the clients of DBMS. This information will also be made available on DBMS portal.
- If the status of the person is under treatment, then the DBMS portal will inform the Corona Control Cell to track location of these persons only. Thus, DBMS portal will provide the information of currently active Corona patients to Corona Control Cell on a daily basis for tracking.
- DBMS portal will also provide the information of currently active Corona patients to public transport (bus, train, flight) and personal transport (car, van, etc.), travel agencies on a daily basis. This will help the travel agencies to check whether any Corona positive patient has booked the ticket or not. If it is booked by any such person, immediately his/her ticket will get canceled. Such cases will also be informed to DBMS portal in the form of feedback.
- Staff of Corona Control Cell will track the currently active Corona patient on a daily basis. If any patient will break the quarantine rule or run from the hospital, Corona Control Cell will immediately inform to the police department.
- Police department will take the information of such patients who broke the quarantine rule from DBMS portal. Police department will also have the facility to track the mobile phones of these persons through DBMS portal.
- Police department will catch such persons immediately. They will re-quarantine these patients. Additionally, they will either give warning to these patients or may take any disciplinary actions on a case-to-case basis.

In this way, it is expected that there will be a smooth flow of information and efficient information exchange will take place. This will improve the functionality of DBMS portal.

9.2.4　CREATION OF DATABASE WITH MICROSOFT SQL SERVER

In this section, database creation is explained with the aid of Microsoft SQL Server 2019 version. For efficient performance of DBMS, it is suggested that one database should be created for a one particular city. For this purpose, a database named 'Corona_Control_DBMS' is created in Microsoft SQL Server. Two tables are created in this database.

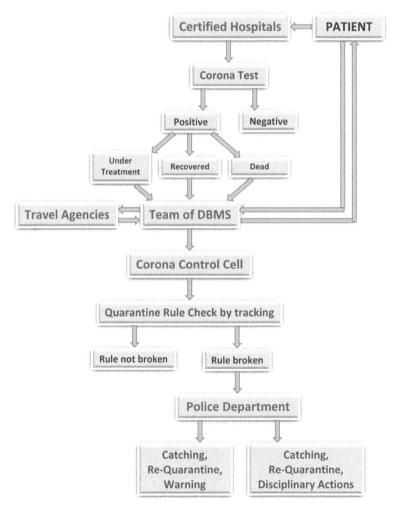

FIGURE 9.5　Schematics of information flow.

Figure 9.6 shows the screenshot of the first table created in the database which stores the information of all the patients undergoing treatment in the hospitals of the city. The name of the table is 'General_Patient_Info' because these may be Corona or non-Corona patients. So, corona test of patients experiencing any of the corona symptoms will be carried out by hospitals or clinics. The result of Corona test (positive or negative) is displayed in the last column of this table.

	Government_ID_Number	Full_Name	Gender	Age	Date_of_Birth	Mobile_Num_1	Mobile_Num_2
1	29084671	Rakhi Singh	Female	48	1971-03-22	490321905	832958492
2	37894489	Adam Fedricks	Male	23	1997-05-18	190380047	497849563
3	60905631	Rohan Sorte	Male	32	1988-01-14	359856792	567930264
4	90478941	Eva James	Female	40	1980-08-25	907321678	310958249

Email_ID	Home_Address	Corona_Test_Result
rakhisingh@gmail.com	4, Revanta House, Vile-Parle, Mumbai	Positive
adam.fed123@gmail.com	904, Fem Apartment, Churchgate, Mumbai	Positive
rohan_sorte@gmail.com	4, Sunshine Bunglow, Sion, Mumbai	Positive
evaj@gmail.com	305, Rose Apartment, Mulund, Mumbai	Negative

FIGURE 9.6   First table created in the database named as 'General_Patient_Info' to store the information of all types of patients (both corona and non-corona).

Figure 9.7 shows the screenshot of second table created in the database which stores the information of the Corona affected patients and named as 'Corona_Patient_Info.' This table is a subset of table of general patient information (first table) as shown in Figure 9.6. Only those patients are included in the second table (Figure 9.7) whose Corona test result is positive as mentioned in the last column of the first table (Figure 9.6). The types of information stored for Corona positive patients is illustrated in the second table (Figure 9.7).

	Government_ID_Number	Full_Name	Gender	Age	Date_of_Birth	Mobile_Num_1	Mobile_Num_2	Email_ID
1	29084671	Rakhi Singh	Female	48	1971-03-22	490321905	832958492	rakhisingh@gmail.com
2	37894489	Adam Fedricks	Male	23	1997-05-18	190380047	497849563	adam.fed123@gmail.com
3	60905631	Rohan Sorte	Male	32	1988-01-14	359856792	567930264	rohan_sorte@gmail.com

Home_Address	Name_and_Address_of_Hospital	Current_Status_of_Patient	Quarantine_Address
4, Revanta House, Vile-Parle, Mumbai	A-7 Hospital, Worli, Mumbai	dead	Not Applicable
904, Fem Apartment, Churchgate, Mumbai	City Centre Hospital, Andheri, Mumbai	under treatment	Home
4, Sunshine Bunglow, Sion, Mumbai	Alexis Hospital, Santacruz, Mumbai	recovered	Not Applicable

FIGURE 9.7   Second table created in the database named as 'Corona_Patient_Info' to store the information of corona positive patients.

The types of information depicted in the first and second table along with their significance are explained as follows:

1. **Government_ID_Number:** It is a unique identification number allotted to each and every citizen of the country. It is also called the National Identification number issued by many countries to their citizens [32]. In India, it is called 'AADHAR' number issued by "Unique Identification Authority of India," which is a government agency [33]. As this is a unique number, it is kept as primary key (attribute) in both the aforementioned tables. The data type is kept as integer. This value cannot be empty while entering details of a person. Hence, 'not null' constraint is applied while defining the table.

2. **Full_Name:** It includes first name, last name, and middle name (if any) of the person. The full name of the patient should be entered because the first name and last name of few persons may be the same. Additionally, full name must be as printed on Government_ID card or National Identification Card. It will be helpful to identify the person easily. Hence, 'not null' and 'unique' constraints are applied while defining the table.

3. **Gender:** It is essential to enter gender in the database. Hence, 'not null' constraint is applied while defining the table. The gender may be male, female or transgender. This is required to segregate the data of patients on the basis of gender. Additionally, if any patient is breaking the rule of quarantine, then based on gender, police staff can be sent to catch the patient. For example, if the patient is female, then few female police staff is must along with male police staff to catch the female patient.

4. **Age:** Age of the person is an important criterion for analysis study of affected patients. It can be statistical analysis carried out by experts in team of DBMS to predict which age group of people are affected more by corona. This study will be helpful from future research perspectives. However, the age of the person cannot be zero or less than zero. Hence 'check (age > 0)' constraint is applied while defining the table.

5. **Date_of_Birth:** Date of birth of the person is needed to calculate the exact age of the person in terms of years, months, and days. For example, if the present age of the person is 23 years, 08 months,

and 20 days on a particular present date. It is calculated only with the help of date of birth.

6. **Mobile_Num_1:** This attribute indicates the primary mobile number of the patient. This can be the mobile number of the patient used for social networking applications such as 'WhatsApp' or any other. It may be the number used for different payment methods such as internet banking, credit card or debit card, etc. The main motive of entering the mobile number in DBMS portal is to track the Corona patient in case he/she breaks the quarantine rule. Additionally, mobile number is required to call or message the patient by team of DBMS, Corona Control Cell, or police department. Thus, information of primary mobile number cannot be empty owing to its high importance. Additionally, it must be unique. Hence, 'not null' and 'unique' constraints are applied while defining the table.

7. **Mobile_Num_2:** Generally, people use dual sim card mobile phones. So, people also use two mobile numbers owing to the issue of calling or internet network connectivity of one sim card in certain geographical locations. Also, it is beneficial for Corona Control Cell and the police department to have two numbers of patient for the purpose of tracking. If one number has network connectivity from the service providers in certain regions, the person can be tracked from other mobile number. Additionally, it must be unique. Hence, 'unique' constraint is applied while defining the table. However, it is not mandatory that a person should use two mobile numbers. Hence, this attribute can be empty if the second mobile number of the patient is not available. Hence, there can be null or no value for this attribute.

8. **Email_ID:** E-mail address is unique place to send detailed instructions along with photographs, text document, presentations, pdf files, etc. E-mail address of the patient is essential for hospital to send different types of reports to patients such as blood scan, X-ray, Corona Test, CT-scan, prescriptions of medicines, etc. Contacting to the patient via e-mail address is considered as an official mode of communication. It is easy to retrieve past e-mails from the inbox of the e-mail address. E-mail address are also provided with the cloud storage system. So, all the patient related documents can be shared to this cloud storage system through

patient's e-mail address. Additionally, it must be unique. Hence, 'unique' constraint is applied while defining the table.

9. **Home_Address:** Home address corresponds to the permanent location of the patient. In case of beds are not available in hospitals for mild affected Corona patients, then such patients are suggested to undergo home quarantine. Corona Control Cell can track home quarantined Corona patients based on this address. Additionally, in case of any emergency, such as the case of the patient is serious, then the patient can inform to Corona Control Cell which will inform to team of DBMS. Team of DBMS will handle these serious issues immediately and hospitals will be informed about case of patient. Patient's personal and contact details along with home address will provided to client hospitals immediately through DBMS portal. Then, nearby hospital or hospital with availability of beds will accept this case. It will be informed to a team of DBMS. Then, hospital will send an ambulance quickly to patient's home address. In this way, home address is a key factor to save the lives of Corona patients.

10. **Corona_Test_Result:** This attribute is only listed in the first table as shown in Figure 9.6. Whenever a patient with symptoms similar to Corona visits a clinic or hospital, Corona test is suggested. If the result of the Corona test is positive, then such patients are treated separately. Whether the result of Corona Test is positive or negative, information of such patients is stored in the first table which represents general patient information. This will be helpful to contact both Corona positive and negative patients in the future. Details of Corona positive patients are separately stored in the second table of the database with few more attributes.

11. **Name_and_Address_of_Hospital:** This information is exclusively stored in the second table of the database (Figure 9.7). The purpose of this information is that the team of DBMS must know treating hospital of each and every Corona patient. This information then will be passed to Corona Control Cell. So, they can directly find out whether the patient is breaking the rule of quarantine or not if the Corona patient is currently being treated in hospitals.

12. **Current_Status_of_Patient:** This information is exclusively stored in the second table of the database (Figure 9.7). There can be

three types of status of any Corona patient. For a particular patient, anyone status is applicable, i.e., (i) the patient is under treatment or (ii) the patient is recovered or (iii) the patient is dead due to Corona infection. So, for a particular Corona patient, this status can be entered in brief as 'under treatment,' 'recovered' or 'dead.' The patients under treatment are the currently active (positive) patients. Information of these currently active patients collected from hospitals will be passed by team of DBMS to different clients such as Corona Control Cell, police department and travel agencies. The information of recovered patients will be removed later from the second table (Figure 9.7) and will be kept only in the first table (Figure 9.6). Additionally, the information of dead patients will also be removed later from the second table (Figure 9.7). It will be kept with the team of DBMS only and will not be shared to other clients. This will only be kept to inform the government about the dead count of Corona affected patients.

13. **Quarantine_Address:** This information is exclusively stored in the second table of the database (Figure 9.7). The under-treatment patient may be either quarantined in hospital or home or any particular public quarantine place allotted by City Municipal Corporation. So, it is very much essential information for Corona Control Cell to track the Corona patients and to check whether they are breaking the rule of quarantine or not. As it is understood that, the status of quarantine will not be applicable to dead or recovered Corona patients. Hence, 'Not Applicable' can be entered in the database as shown in Figure 9.7.

The commands of structured query language (SQL) to create and use the database are as follows:

create database Corona_Control_DBMS;
use Corona_Control_DBMS;

The commands of SQL to create the first table in the database (Figure 9.6) are as follows:

CREATE TABLE General_Patient_Info
(Government_ID_Number INT NOT NULL PRIMARY KEY,

Full_Name VARCHAR(100) NOT NULL UNIQUE,
Gender VARCHAR(10) NOT NULL,
Age INT CHECK(age > 0),
Date_of_Birth DATE,
Mobile_Num_1 INT NOT NULL UNIQUE,
Mobile_Num_2 INT UNIQUE,
Email_ID VARCHAR(100) UNIQUE,
Home_Address VARCHAR(300) NOT NULL,
Corona_Test_Result VARCHAR(10) NOT NULL);

The commands of SQL to insert the information in the rows of first table in database (Figure 9.6) are as follows:

Note: Dummy (non-real) information is entered below:

INSERT INTO General_Patient_Info VALUES('37894489,' 'Adam Fedricks,' 'Male,' '23,' '1997-05-18,' '190380047,' '497849563,''adam.fed123@gmail.com,' '904, Fem Apartment, Churchgate, Mumbai,' 'Positive');

INSERT INTO General_Patient_Info VALUES('90478941,' 'Eva James,' 'Female,' '40,' '1980-08-25,' '907321678,' '310958249,''evaj@gmail.com,' '305, Rose Apartment, Mulund, Mumbai,' 'Negative');

INSERT INTO General_Patient_Info VALUES('60905631,' 'Rohan Sorte,' 'Male,' '32,' '1988-01-14,' '359856792,' '567930264,''rohan_sorte@gmail.com,' '4, Sunshine Bunglow, Sion, Mumbai,' 'Positive');

INSERT INTO General_Patient_Info VALUES('29084671,' 'Rakhi Singh,' 'Female,' '48,' '1971-03-22,' '490321905,' '832958492,''rakhis-ingh@gmail.com,' '4, Revanta House, Vile-Parle, Mumbai,' 'Positive');

The command of SQL to display or retrieve the first table in database (Figure 9.6) is as follows:

select * from General_Patient_Info;

The commands of SQL to create the second table in the database (Figure 9.7) are as follows:

CREATE TABLE Corona_Patient_Info
(Government_ID_Number INT NOT NULL PRIMARY KEY,
Full_Name VARCHAR(100) NOT NULL UNIQUE,

Gender VARCHAR(10) NOT NULL,
Age INT CHECK(age > 0),
Date_of_Birth DATE,
Mobile_Num_1 INT NOT NULL UNIQUE,
Mobile_Num_2 INT UNIQUE,
Email_ID VARCHAR(100) UNIQUE,
Home_Address VARCHAR(300) NOT NULL,
Name_and_Address_of_Hospital VARCHAR(100) NOT NULL,
Current_Status_of_Patient VARCHAR(20) NOT NULL,
Quarantine_Address VARCHAR(300) NOT NULL);

The commands of SQL to insert the information in the rows of the second table in database (Figure 9.7) are as follows:

Note: Dummy (non-real) information is entered below. These are the same patients as shown first table (Figure 9.6) but only Corona positive patients.

INSERT INTO Corona_Patient_Info VALUES('37894489,' 'Adam Fedricks,' 'Male,' '23,' '1997-05-18,' '190380047,' '497849563,''adam. fed123@gmail.com,' '904, Fem Apartment, Churchgate, Mumbai,' 'City Center Hospital, Andheri, Mumbai,' 'under treatment,' 'Home');

INSERT INTO Corona_Patient_Info VALUES('60905631,' 'Rohan Sorte,' 'Male,' '32,' '1988-01-14,' '359856792,' '567930264,''rohan_ sorte@gmail.com,' '4, Sunshine Bungalow, Sion, Mumbai,' 'Alexis Hospital, Santacruz, Mumbai,' 'recovered,' 'Not Applicable');

INSERT INTO Corona_Patient_Info VALUES('29084671,' 'Rakhi Singh,' 'Female,' '48,' '1971-03-22,' '490321905,' '832958492,''rakh-isingh@gmail.com,' '4, Revanta House, Vile-Parle, Mumbai,' 'A-7 Hospital, Worli, Mumbai,' 'dead,' 'Not Applicable');

The command of SQL to display or retrieve the second table in database (Figure 9.7) is as follows:

select * from Corona_Patient_Info;

9.3   RESULTS AND DISCUSSION

Figure 9.8 shows the database created with Microsoft SQL Server. Left elongated section shows hierarchy of database, tables, and attributes. Central upper half section shows the SQL query writing portal. While, central lower half section displays requested table information.

This database needs to be operated with a software system to update, add, delete, and retrieve the information. This will then form the complete DBMS portal. With due permission of Government authorities, it can be employed in Corona affected cities to break the chain of Corona spread. A selected team of professionals will make it a role model system than any other preventive measures.

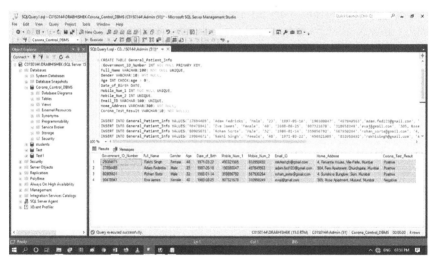

FIGURE 9.8   Database created with Microsoft SQL Server.

9.4   CONCLUSIONS

The highlights of proposed DBMS portal are as follows:

- The proposed system consists of governing bodies such as team of DBMS, certified hospitals, Corona Control Cell, police department, and travel agencies.
- Tracking the Corona patients and quarantine them after catching for any quarantine rule break will definitely restrict the corona spread.

- Creation of a table of general patient information in the database will help to store basic details of patients, which will be useful for future reference.
- Creation of table of Corona patient information which is a subset of table of general patient information in the database will help to get all important information of Corona patient. It will be helpful to track the people to check whether they are obeying the rule of quarantine or not.
- This proposed system will be helpful to provide emergency services to patients through DBMS clients such as Corona Control Cell, police department, and hospitals with the help of information provided by team of DBMS.
- The travel agencies will refrain the currently active Corona patients to travel outside the city by accessing their details through DBMS portal.
- With the due permission of Central or State Government authorities, it can be employed in a variety of cities.
- Team of skilled and trained professionals can make it as a system of prime importance.
- In the future, more necessary changes can be implemented in database handling system (DBMS portal) if any issues faced during practical application in the society.

The database is created with the help of Microsoft SQL Server. This database needs efficient software platform to form the complete DBMS portal. This system should be first operated in any one city on a trial basis. After successful implementation practice, it can be launched globally to effectively control the spread of Corona Virus. The Government of any country can enhance the performance of this system by employing highly skilled and trained professionals.

KEYWORDS

- **coronavirus**
- **database**
- **database management system**
- **DBMS portal**
- **Microsoft SQL server**
- **structured query language (SQL)**

REFERENCES

1. Yousefpour, A., Jahanshahi, H., & Bekiros, S., (2020). Optimal policies for control of the novel coronavirus (COVID-19). *Chaos Solitons Fractals., 136*, 109883.
2. *Coronavirus*, (2020). World Health Organization. [online]. Available: https://www.who.int/health-topics/coronavirus#tab=tab_3 (accessed on 22 June 2021).
3. Wu, F., Zhao, S., Yu, B., Chen, Y. M., Wang, W., Song, S. G., et al., (2020). A new coronavirus associated with human respiratory disease in China. *Nature., 579*, 265–269.
4. Zhou, P., Yang, X. L., Wang, X. G., Hu, B., Zhang, L., Zhang, W., et al., (2020). A pneumonia outbreak associated with a new coronavirus of probable bat origin. *Nature, 579*, 270–273.
5. Kang, S., Peng, W., Zhu, Y., Lu, S., Zhou, M., Lin, W., Wu, W., et al., (2020). Recent progress in understanding novel coronavirus associated with human respiratory disease: Detection, mechanism and treatment. *Int. J. Antimicrob. Ag., 55*(5), 105950.
6. Carlos, W. G., Dela, C. C. S., Cao, B., Pasnick, S., & Jamil, S., (2020). Novel Wuhan (2019-nCoV) coronavirus. *Am. J. Respir. Crit. Care Med., 201*(4), 7, 8.
7. *Coronavirus Worldwide Graphs*, (2020). Worldometer. [online]. Available: https://www.worldometers.info/coronavirus/worldwide-graphs/ (accessed on 22 June 2021).
8. *Coronavirus and COVID-19: What You Should Know*, (2020). WebMD. [online]. Available: https://www.webmd.com/lung/coronavirus#3-5 (accessed on 22 June 2021).
9. Van, D. N., Bushmaker, T., Morris, D. H., Holbrook, M. G., Gamble, A., Williamson, B. N., & Lloyd-Smith, J. O., (2020). Aerosol and surface stability of SARS-CoV-2 as compared with SARS-CoV-1. *New Engl. J. Med., 382*(16), 1564–1567.
10. Suman, R., Javaid, M., Haleem, A., Vaishya, R., Bahl, S., & Nandan, D., (2020). Sustainability of Coronavirus on different surfaces. *J. Clin. Exp. Hepatology, 10*(4), 386–390.
11. *One Graphic Shows How Long the Coronavirus Lives on Surfaces Like Cardboard, Plastic, and Steel*, (2020). Business Insider India. [online]. Available: https://www.businessinsider.in/science/news/one-graphic-shows-how-long-the-coronavirus-lives-on-surfaces-like-cardboard-plastic-and-steel/articleshow/74720814.cms (accessed on 22 June 2021).
12. *Considerations for Public Health and Social Measures in the Workplace in the Context of COVID-19*, (2020). World Health Organization. [online]. Available: https://www.who.int/publications/i/item/considerations-for-public-health-and-social-measures-in-the-workplace-in-the-context-of-covid-19 (accessed on 22 June 2021).
13. Leung, N. H. L., Chu, D. K. W., Shiu, E. Y. C., et al., (2020). Respiratory virus shedding in exhaled breath and efficacy of face masks. *Nat. Med., 26*, 676–680.
14. Cheng, V. C., Wong, S. C., Chuang, V. W., So, S. Y., et. al., (2020). The role of community-wide wearing of face mask for control of coronavirus disease 2019 (COVID-19) epidemic due to SARS-CoV-2. *J. Infection, 81*(1), 107–114.
15. Eikenberry, S. E., Mancuso, M., Iboi, E., Phan, T., Eikenberry, K., Kuang, Y., Kostelich, E., & Gumel, A. B., (2020). To mask or not to mask: Modeling the

potential for face mask use by the general public to curtail the COVID-19 pandemic. *Infec. Dis. Model., 5*, 293–308.

16. *Advice on the Use of Masks in the Context of COVID-19: Interim Guidance*, (2020). World Health Organization. [online]. Available: https://apps.who.int/iris/handle/10665/331693 (accessed on 22 June 2021).

17. *N95 Respirators, Surgical Masks, and Face Masks*, (2020). FDA (Food and Drug Administration). [online]. Available: https://www.fda.gov/medical-devices/personal-protective-equipment-infection-control/n95-respirators-surgical-masks-and-face-masks (accessed on 22 June 2021).

18. *How Effective Are Masks and Other Facial Coverings at Stopping Coronavirus?* (2020). Laborers' Health & Safety Fund of North America. [online]. Available: https://www.lhsfna.org/index.cfm/lifelines/may-2020/how-effective-are-masks-and-other-facial-coverings-at-stopping-coronavirus/ (accessed on 22 June 2021).

19. What to Know About KN95 Face Masks, (2020). *The Wall Street Journal.* [online]. Available: https://www.wsj.com/articles/what-to-know-about-kn95-face-masks-11596460569 (accessed on 22 June 2021).

20. Qian, Y., Willeke, K., Grinshpun, S. A., Donnelly, J., & Coffey, C. C., (1998). Performance of N95 respirators: Filtration efficiency for airborne microbial and inert particles. *Am. Ind. Hyg. Assoc. J., 59*(2), 128–132.

21. Wong, J., Goh, Q. Y., Tan, Z., Lie, S. A., Tay, Y. C., Ng, S. Y., & Soh, C. R., (2020). Preparing for a COVID-19 pandemic: A review of operating room outbreak response measures in a large tertiary hospital in Singapore. *Can. J. Anesth., 67*(6), 732–745.

22. *COVID-19 Hierarchy of Controls*, (2020). Cornell University, New York, USA. [Online]. Available: https://ehs.cornell.edu/campus-health-safety/occupational-health/covid-19/covid-19-hierarchy-controls (accessed on 22 June 2021).

23. *Hierarchy of Controls*, (2015). Centers for Disease Control and Prevention. [online]. Available: https://www.cdc.gov/niosh/topics/hierarchy/default.html (accessed on 22 June 2021).

24. *List N: Disinfectants for Use Against SARS-CoV-2 (COVID-19)*, (2020). Environmental Protection Agency, United States. [online]. Available: https://www.epa.gov/pesticide-registration/list-n-disinfectants-use-against-sars-cov-2-covid-19 (accessed on 22 June 2021).

25. Wang, Y., Liu, J., He, X., & Wang, B., (2018). Design and realization of rock salt gas storage database management system based on SQL Server. *Petroleum., 4*(4), 466–472.

26. Yue, L., (2018). The comparison of storage space of the two methods used in SQL server to remove duplicate records. *Procedia Comput. Sci., 131*, 691–698.

27. *Applications of DBMS*, (2019). Tutorial and Example. [online]. Available: https://www.tutorialandexample.com/applications-of-dbms/ (accessed on 22 June 2021).

28. Lo, E., Binnig, C., Kossmann, D., Özsu, M. T., & Hon, W. K., (2010). A framework for testing DBMS features. *The VLDB J., 19*(2), 203–230.

29. *Components of DBMS*. Study Tonight. [online]. Available: https://www.studytonight.com/dbms/components-of-dbms.php (accessed on 22 June 2021).

30. *DBMS Database Models*. Studytonight. [online]. Available: https://www.studytonight.com/dbms/database-model.php (accessed on 22 June 2021).

31. *Best Relational Databases Software*, (2020). G2. [online]. Available: https://www.g2.com/categories/relational-databases (accessed on 22 June 2021).
32. *National Identification Number*. Wikipedia. [online]. Available: https://en.wikipedia.org/wiki/National_identification_number (accessed on 22 June 2021).
33. *My Aadhaar*, (2016). Unique Identification Authority of India, Government of India. [online]. Available: https://uidai.gov.in/ (accessed on 22 June 2021).

CHAPTER 10

Wi-Fi-Based Proximity Social Distancing Alert to Fight Against COVID-19

MAYURI DIWAKAR KULKARNI and KHALID ALFATMI

Assistant Professor, Department of Computer Engineering,
SVKM's Institute of Technology, Dhule, Maharashtra, India,
E-mail: mayuridkulkarni@gmail.com (M. D. Kulkarni)

ABSTRACT

The Pandemic COVID-19 outbreak is only due to the transmission of the virus. This virus transmission is mainly possible due to coughing, sneezing, or droplets. Hence the WHO issued the guideline as to the use of masks and maintain social distancing.

Social distancing is used to break the chain of transmission. The social distancing parameter is decided as a 1.5-meter minimum in case of having a mask. So that after sneezing, coughing the droplets may translate to less than a 1.5-meter area only. Due to this, the transmission chain of the COVID-19 will be going to shrink.

The one more major issue faced by the Government is to trace the person who was in contact with the COVID positive. There is no such technical solution found. Due to this, the chain of transmission is not possible to break.

By considering this factor, the proposed technique will be suggesting Wi-Fi-based proximity estimators, which will continuously monitor the Wi-Fi-enabled devices. Wifi-enabled devices such as smartphones will specify the position of another person whose Wi-Fi is turned on. If that distance is more than 2 meters, it will not alert. But if the distance is less than 2 meters, then it will go to alert. Also, storing the details of the person will be in proximity less than 1.5 meters.

This proposed technique not only maintains the social distance but also useful to provide the contact tracing.

10.1   INTRODUCTION

Pandemic is generally as "an epidemic present worldwide or widely covered in the area crossing the nations, continent, and affecting a large number of people" [1]. The very first evidence of the severe acute respiratory syndrome coronavirus 2 (SAR-CoV-2) virus in Wuhan, China. The major origin of SAR-COV-2 is the bat. This infection begins with infecting the lungs. In the beginning, it is classified as pneumonia of unknown etiology. After investigating the Chinese Center for Disease Control (CDC) and Prevention reached to the conclusion that the virus belongs to corona [33].

The symptoms of this disease are fever, coughing, headache, body pain, etc. The coronavirus disease 2019 (COVID-19) spread due to direct, indirect, or close contact with infected people. These spreads due to mouth or nose secretions which include saliva, respiratory, or droplet secretion [2]. The person who has close contact with infected droplets within 1 meter gets into the mouth, eyes or nose may lead to getting infected [6]. The proportion of cases tested positive due to the SAR-COV-2 virus concerning asymptomatic cases is 16%.

The transmission of the virus may be at various places. These places are listed below:

- Indoor spaces such as offices, restaurants, meetings, sharing, etc.;
- Lack of social distancing at the canteen, dressing rooms, transport, and accommodation;
- Close contact with COVID-19 cases.

There are various ways of virus transmission such as Contact, Direct, Indirect, Droplet, Aerosol. Among these reasons, the spread of COVID-19 virus infection can be through contact, droplets, aerosol. The transmission of the virus may be in direct or indirect contact with the person in proximity. The droplets may travel in proximity after coughing, sneezing, etc., within the short range. Also, these droplets may be present at any place in an environment. This may lead to contamination of the place. And by touching such a contaminated surface, the chain of transmission gets continued, this lead to the indirect transmission of the virus. Table 10.1 shows the mode of person-to-person transmission of respiratory viruses [35].

TABLE 10.1   Modes of Virus Transmission

Transmission Type	Transmission Modes
Contact transmission	Direct and indirect mode
Direct transmission	Transmission of the virus from one infected person to another person.
Indirect transmission	Transmission virus due to intermediate objects which were in contact with an infected person.
Droplet transmission	Transmission due to droplets in the air while sneezing, coughing.
Aerosol transmission	Virus transmission through the air.

In the case of indirect virus transmission, the virus is active on the different surfaces from few hours to days. Table 10.2 indicates the lifespan of the virus on surfaces such as metal, plastic, wood, air [2, 10]. This may also lead to one of the reasons for the transmission of the virus.

TABLE 10.2   Lifespan of SAR-COV-2 [2, 10]

SL. No.	Surface Material	Lifespan of SAR-COV-2
1.	Airborne	3 hours
2.	Copper	4 hours
3.	Paper	24 hours
4.	Wood	4 days
5.	Plastic, Steel	2–3 days
6.	Silver, Iron, Gold	5 days
7.	Iron	5 days

Droplet, airborne are the transmission leading confusion over the physical behavior of the virus particles. In case of such virus, particles are inhaled through breathing in aerosols. This can possible at long range, it is similar to when close to someone [39].

Among all these transmission reasons, the key reason for the transmission of the virus is contact between people which may be directly or indirectly. Contact with infected people leads to the possibility of transmission at a higher rate [5]. According to the CDC and Prevention, close contact timing of approximately 15 minutes with an infected person may lead toward the transmission of the virus [38]. This transmission of the virus may be due to the insufficient ventilation and more contact hours following the norms of the social distancing. The general example of it may be family members sharing the same home, roommates in a hostel, co-workers sharing the same work area.

Depend upon the type of exposure and duration of the contact is important to check the boundaries of getting the risk of infection. The risk of transmission has majored as high risk, medium high risk, medium risk, and low risk. These risk factors are based on the following aspects such as type of exposure, and the examples are listed for the same, shown in Table 10.3.

TABLE 10.3   Risk Factor and Exposure with Examples [36]

Risk	Type of Exposure	Examples
High risk	Living in the same household or the same room without wearing a mask with an infected person without recommended personal protective equipment (PPE), or facemask	Domestic partner, Healthcare personnel
Medium-high risk	Frequent contact with an infected person or performing vital signs and phlebotomy on masked patients without recommended PPE.	Close work associate, family members visited prolonged
Medium risk contact	Close contact with an infected person and not having exposures that meet to the medium-high risk and high-risk definition	Colleagues who work less closely together but regular face to face contact
	Without recommended PPE in contact with an infected person for more than 2 minutes.	
Low-risk contact	In the same indoor environment with or within 2 hours, contact with an infected person for 1–2 minutes involving 1–2 minutes direct or indirect contact with recommended PPE	Shared hospital or outpatient waiting room or entered space

The one infected person comes in contact with 2 people in a day then the possibility of transmission of the virus is 2. Now, these 2 people individually come in direct contact with 2 people then the possibility of transmission is 4. From this, we can analyze the transmission rate by Table 10.4. Table 10.4 indicates that a single infected person comes in contact with only 2 people on a single day, and such sequence is continued by the people who come in proximity to an infected person then within 5 days, 32 infectious people will be probable may occur.

TABLE 10.4   Transmission of SAR-COV-2

Transmission Rate	Number of Infected People		
Day	2	3	4
Day 1	2	3	4
Day 2	4	9	16
Day 3	8	27	64
Day 4	16	81	256
Day 5	32	729	1024

The formula for the transmission of the infection is given as follows:

$$I = R^D$$

where; I is the number of infected people; R is the rate of transmission, and d is the number of days.

The contact rate is the rate that indicates the rate of transmission. This indicates the number of persons comes in proximity and the risk associated with it. The reproduction number is an indicator of disease transmission.

Figure 10.1 shows that one infected person can transmit the infection if only 2 people in a vicinity spread of the disease within 3 days will have a minimum of 8 people. If this transmission chain can be broken down due to some technique. So that the transmission chain will be broken only by keeping a safe distance. Table 10.4 shows the statistics indicate if the transmission rate varies from 2 to 3 then raise in a count on the 5th day will be 629 and leads to a rise in infection due to COVID-19.

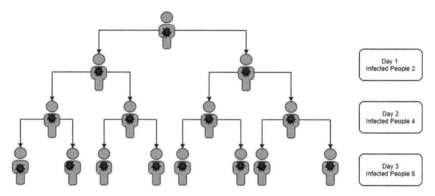

FIGURE 10.1   Transmission of the virus from 1 infected with Ro = 2.

Based on the chain of transmission the transfer of disease is broadly categories into the following ways [4, 6]:

- Community transmission;
- Cluster transmission;
- Interrupted transmission.

In the case of community transmission, the source of infection is not at all traceable and this is also known as stage 3 in COVID-19. In this case, the transmission of the virus is not traceable where the maximum number of people is infected [4, 6].

In the case of Cluster transmission, the source of transmission can be traceable. Cluster transmission can be minimized by using preventive measures such as maintaining social distance. By keeping infected people in isolation phase. The people who are in a vicinity area of the infected people can be in quarantine so that the chain of transmission breaks. Table 10.5, shows the cumulative cases of infected due to these two types of transmission such as Community transmission and Cluster transmission on 14 September 2020 [6].

Interrupted transmission in which the transmission is interrupted. Due to this, the occurrences of newly infected cases are very less [6].

Transmission is not only due to the symptomatic patients but also due to asymptomatic patients [6]. Such cases are observed in many countries. These asymptomatic patients come in proximity to the uninfected person then it may lead to the transmission of the SAR-Cov-2 virus. To avoid this transmission, the need for proximity barrier is much more important. These asymptomatic patients may act as a transmitter. So, the main focus in both these cases of transmission is only to maintain social distancing. Due to this, it will break the transmission chain.

TABLE 10.5 Infected Cases Up to 14 September 2020 based on Classification of Transmission [4]

Name of Country	Cumulative Cases	Transmission Type
United State	6,426,958	Community
India	4,846,427	Cluster
Brazil	4,315,687	Community
Russia	1,068,320	Cluster
Peru	722,832	Community
Colombia	708,964	Community
Mexico	663,973	Community
South Africa	649,793	Community
Spain	566,326	Cluster
Argentina	546481	Community

There are little shreds of evidence of airborne transmission of SAR-CoV-2 is also available. This airborne transmission may get reported in indoor environments. It has a short range of transmission due to small droplets present in the air. These cases of transmission are reported at Washington, Guangzhou, and Huan. In Guangzhou, China reported 10 cases in January 2020 [5].

Figure 10.2 shows the number of cases up to 14 September 2020, due to symptomatic or asymptomatic infection of the virus all over the World. This detailed data is available at the European Center for Disease Prevention and Control (ECDPC) [3].

10.2   WHO GUIDELINES TO FIGHT WITH COVID-19

The guideline issued by WHO to prevent the spread of COVID-19 is by maintaining social distancing at social gatherings such as schools, colleges, workplaces, and religious prayer centers by keeping a safe distance and wearing masks [6, 7]. If the social distancing parameter is neglected by the people, then the spread of COVID-19 will woefully in a community. This may require extremely well-equipped hospitals. Due to this World Health Organization (WHO) urges people to maintain social distancing along with hygiene and sanitization. The sanitization of the public places and workplaces after some time interval is essential along with the personal sanitization maintenance.

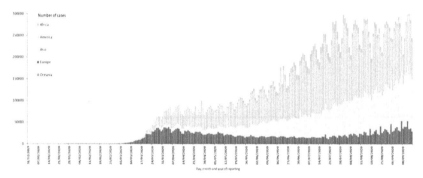

FIGURE 10.2   Worldwide distribution graph of number of cases up to 14 September 2020 [3].

While disclosing the interim guidelines, the main focus was on social distancing with a minimum of 1.5 meters along with the mask.

Figure 10.3 shows the case 1, where the safe distance of 2 meters is maintained along with wearing a mask by two persons which cover the mouth and nose. In case 2, both persons maintain a safe distance, but they do not wear the mask.

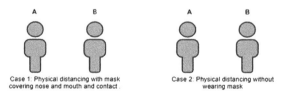

Case 1: Physical distancing with mask covering nose and mouth and contact .

Case 2: Physical distancing without wearing mask

FIGURE 10.3   Scenarios considering physical distancing.

Now in the case of case 1, the persons wore masks and maintained social distance hence the probability of physical contact is minimized.

Suppose any of the person sneezing, coughing then transmission of droplets is very lesser. Due to this spread of the virus may be reduced.

In Case 2, the distance is safe, but any amongst these two people if any infected person sneeze, cough then droplets more than 5–10 micrometers in a length can be exposed to 100–1000 virus particles which lead to a probability of infection [34].

To control the spread of COVID-19 early detection of the suspected cases and testing them. After test isolating such cases and this isolation is still the case is not tested negative. Then the major role is to trace the contact. After tracing the contacts again, testing the suspect and if the suspect is negative then quarantine for 14 days [7]. An infected person starts the transmission of the virus from 2 days after tested positive or after the symptoms are recognized. The period of symptoms diagnosis may be at the Second, Fifth, or fourteenth day. Hence the quarantine time was decided as 14 days [7]. The symptoms development is estimated at 14 days is 5% [31].

During this quarantine period if the person tested positive then isolated or else can execute the routine. In the initial stage, various countries started the lockdown process to minimize the outbreak. After the lockdown phase, the process of unlocking begins. During the lockdown or unlocking process various guidelines issued by the WHO to prevent the spread of COVID-19. To guidelines were suggested by WHO to prevent the spread of COVID-19 while executing the daily routine. The guidelines for the unlock are such as:

- Early detection of suspected case and test suspected cases and trace contacts;
- Limiting gatherings and reducing mobility;
- Physical distancing at least 1 meter with personal hygiene. Wearing mask properly;
- Safe transportation such as walking, cycling, or limiting the count of people;
- Keeping small groups of teachers and students by maintaining appropriate social distancing at schools [8].

After the unlock the people are joining their workplace. Due to this, the transmission probability of infection from one person to another person may increase. This will lead to a rise in the transmission rate. To minimize transmission needs to reduce the transmission chain of the virus. This can be achieved by proper strategies with the infected people such as isolation besides quarantine to the people in the proximity of the infected person. Maintaining a social distancing minimum of 1.5 meters may reduce the transmission rate of less than 1 indicated the end of the transmission. This leads to no new case of getting infected due to COVID-19.

10.2.1   TECHNICAL APPROACHES USED TO MINIMIZE TRANSMISSION RATE IN PROXIMITY

Proximity measuring technologies:

- Wi-Fi infrastructure will track the location of the devices so that based on this, the proximity can be calculated.
- Bluetooth tracking infrastructure is also helpful to measure proximity from the location of connected devices.
- Wearable devices or smartphone apps can estimate the distance between proximity devices.
- AI-based proximity estimation from video streaming by analyzing the videos.
- GPS for the proximity estimation.
- No-touch access controls such as RFID, Wireless badges can be used. These will be identifying the people entering and leaving zones along with the count of the people in proximity.
- Bluetooth, GPS, Wi-Fi-based proximity measure techniques can easily be implemented because of the availability of smartphones. And these all the sensors are present in it. By considering this, different countries used different technology platforms after the lockdown phase. This lockdown phase can be said preparation time for all the countries to come up with technical solutions.

10.2.2   THE TECHNOLOGIES USED BY THE DIFFERENT GOVERNMENTS TO CONTROL OUTBURST

10.2.2.1   SINGAPORE

Singapore launched a phone application that exchanges Bluetooth signals when individuals are in proximity to each other. This keeps the details of those people who were in proximity for 21 Days. This identifies those who may get infected due to the infected person. Due to this reason, Singapore has maintained the lowest per-capita COVID-19 mortality rates in the world. The main focus of Singapore was to minimize the transmission rate. Due to the use of the phone application, they achieved the same [9].

10.2.2.2   SPAIN

In Spain, the I-MOVE-COVID-19 mobile application is to obtain the contacts of the infected patients due to SAR-COV-2. Spanish National Research Council developed a data science project using mobile. This data science project's main aim was to depression of COVID-19 in containment zones. For that, they gathered data from different sources. This gathered data is used to decide the strategies to improve social distancing. Based on those strategies, social distancing guidelines are issued by the authorities to control the outburst COVID-19 [11].

10.2.2.3   INDIA

The Government of India has initiated different applications and different techniques to cope up with the spread of the COVID-19. Due to the dense population in India, it was difficult to gather the data initially and later to implement the social distancing policies. Hence some initiatives are such as Aarogya Setu App to trace the contacts, the Corona Kavach app to use for the tracking, AI-powered Drishti to maintain social distancing in a public place.

- **Aarogya Setu:** This app is launched by the government app which is a contact tracing and syndromic mapping app. This app self-diagnosed 1,50,000 high-risk people and among them, 380,00 people infected because of the COVID-19 [14].
- **Corona Kavach:** This app is launched by the ministry of electronics and information technology (MeitY) in association with the Ministry of Health and Family Welfare (MHFW). The main aim of this app is to track the location of a person and to protect from infection. It has different color schemes to alert such as based on proximity [37]:

- Green for all fine;
- Orange for doctor;
- Yellow for quarantine;
- Red for infected.

- **AI-Powered Drishti:** Defense Institute of Advanced Technology, Pune built a national level product called "DRISHTI (digital real-time artificial intelligence system for social distancing with timely intervention)" to recognize a face, expression, and gesture using the use of recent technologies like artificial intelligence (AI) [12]. This AI-powered Drishti is capable of recognition of real-time images as well as videos. It has three digital eyes among them first eye will detect social distancing, the second eye focuses on face mask related violation, and the third eye is for counting the persons in public places [15]. Due to these features, the Government of India suggested utilizing these in malls, in metros such as Delhi, Mumbai, etc., at public places such as traffic signals, at bus or railway stations.

- **AI-Powered Tools:** AI-Powered Tools are used to combat with COVID-19 in different states of India [13].

- In Kerala, "Thermal and optical Imaging Camera "is used for fever screening. Not only this but also this screening is done using the "Srishti Robotics." This robot is used for delivering food, bulk medicine delivery, and video interactive way of attending patients. Due to this, the transmission of the virus is not possible. Humans are replaced with Robots.

- The state of Karnataka traced people's travel history to foreign. Based on the available data analyzed. Those suspects are needs to install the Corona Watch app. This app keeps a keen eye on travelers. The activities of self-quarantine people are tracked by uploading the photos. These photos must be geotagged. Due to which possibility of transmission of the virus from this self-quarantine to local people can be minimized. This app keeps the details of the people who are self-quarantining. They need to click and upload their photo periodically. This uses the geotags to check that is quarantine or not.

- Uttar Pradesh used Adverb Technologies who assist the Government to deliver food, medicines at quarantine centers without any human involvement. Also used AI-powered crowd surveillance called Staqu. This limits the spread of transmission. Also, a video analytics platform JARVIS is being used by the Government to help for maintaining social distancing. This is helpful to the authorities.

- In Bihar, NIPI used an AI-powered tool to track COVID-19 cases which determine the potential patients through cough sound analysis. This detects the infectious case through cough sound analysis [13].

10.2.3   HOW IMPORTANT CONTACT TRACING APPS IMPORTANT?

As we know that COVID-19 virus transmission is due to the physical contact or transfer of droplets. This transfer of the virus from one person to another is the major reason for occurrences of more positive cases. Due to this, it is very important to know that an infected person was with whom in proximity. That proximity and duration of the particular proximity can easily indicate the probability of virus transmission.

And if such a count is available with the Government, this will be useful to test infection status, and if the person is not tested positive, then they may ask people to quarantine. During that period, more focus will be on infected people and quarantine people [23].

This contact tracing keeps the transmission of the virus in a limitation due to which community transmission can be controlled. Once the community transmission occurred then to control such a situation is not possible to the medical healthcare sector due to the limited infrastructure.

Contact tracing helps to minimize the outbreak of SAR-CoV-2 and due to this, the virus transmission will be only in a cluster. In the case of cluster transmission, the probability of minimizing the transmission rate can lower down by taking some preventive measures or by strictly following the social distancing guidelines. Hence the rate of transmission can be controlled.

10.3   TECHNOLOGY HELP FOR CONTACT TRACING IN INDIA

The Government of India initiated contact tracing in India using the Arogya Setu app [28]. This app is available in 11 regional languages in India such as Hindi, Gujarati, Marathi, Bengali, Panjabi, Odia, Kannada, Tamil, Malayalam, Assamese, Telugu. This Arogya Setu uses contact tracing.

This contact tracing will restore the details of all the people. This restoring of details is only of people who come in contact. If anyone of them tested positive due to COVID-19 then the Arogya Setu app will immediately inform. This also gathers the data vicinity distance [17].

This app uses GPS, Bluetooth, AI, and contact tracing using data analytics [16]. This app keeps the safe by providing features such as first or second-degree contact with COVID-19 positive person, alerts, and gets timely medical checkup help [17].

Contact tracing in the Arogya Setu app is through Bluetooth [24]. Through the Arogya Setu installed in two smartphones in proximity collects the information. If the two people already tested positive, the app will notify and allow the Government to trace contacts [18].

This uses the RSSI to check the proximity and send the information to connected devices if the infected person is in proximity and sends the GPS location of the rest of the connected devices through the GPS to the Government.

10.3.1 DISTANCE MEASURING TECHNIQUES FOR KEEPING EYE ON SOCIAL DISTANCING IN INDIA

Different Technical ways to know the distance between two people by considering technologies such as Image or Video, geotags, global positioning system (GPS) signals, RF signals using received signal strength indication (RSSI). We will be discussing them in detail.

10.3.1.1 IMAGE OR VIDEO

AI-powered DRISTHI uses Images or videos to calculate the distance among the people. This uses image processing algorithms to detect the objects and also check the distance among these objects in a system. This system continuously captures the videos from the CCTV and continuously monitors the crowd. This is always keeping an eye on people who are trying to violate the COVID-19 unlock norms such as maintaining the social distance, wearing a mask, and keeping count in crowded regions. Due to this, its applicability is in public places like airports, railway stations, Bus terminals, or at social gathering spaces.

This can be implemented in the shopping malls where plenty of CCTV cameras are in usage [15]. The real-time face detection is possible using the OpenCV library, Dlib [19].

10.3.1.2   GPS

GPS provides the longitude and latitudes. It is 90% accurate for outdoor positioning. The distance calculation between transmitter and receiver signals using signal propagation time and velocity of the signals. The distance between transmitter and receiver is calculated using the following equation [27]:

$$x_n + 1 = x_n + H^{-1} (t - f(x_n))$$

where; x is a vector of user position; H is the measurement matrix of partial derivatives $H = f(x)$.

GPS technique is used in the Arogya Setu app to trace the contacts. This application user is 15,80,00,000 on 22 September 2020 [29]. This provides a precise estimation of the distance up to 10 to 15-meter outdoors [30].

10.3.1.3   RSSI

To collect information from the wireless sensor network is distributed by the sensor nodes, which are usually in fixed locations. The received signal strength indicator measures the signal strength between connected devices. This applies to an indoor positioning system where the devices are nearby [21]. RSSI ranging is transforming the attenuation of the signal strength. This signal strength is in between transmitting and receiving nodes signal propagation distance by the free space path loss. The free space propagation model is used to predict the receiving signal strength under the condition of signal propagation without shielding and reflector in the transmission path. The distance is calculated using the Friis Transmission equation is given below [26]:

$$Pr/pt = ((Ar\ At)/(d^2\ \lambda^2))$$

where; Pt is the power of transmitting antenna input element; Pr is the power of receiving antenna output element; Ar is the receiving antenna; At is the transmitting antenna; d is a distance between antennas; λ is the wavelength of the radio frequency.

The distance between the unknown node and the base station is d, *reference* node, and the base node is represented with d_0 hence the signal intensity at d, d_0 with n loss path. is a random variable with an average of 0. The factors x and are related to the height of the environment. The logarithmic expression is given below [20]:

$$PL(d) = P(d_0) - 10\eta \log \frac{d}{d_0} + x_\delta$$

The factors are responsible for the error in RSSI because of two reasons such as the influence of the obstruction and multi-path effects as well as the RSSI path loss model.

The positioning principle is constructed mapping the RSSI and distance according to function:

$$\rho = \partial - 10\ \beta\ \log(d)$$

where; ρ is RSS value at the position d with ∂ RSS value at distance β [32]

10.3.2 CONTACT TRACING MECHANISM

The interaction between individuals is one way to trace the contacts. In the case of videos or images, it is very easily possible, but it is a costly system to use. The cost associated with CCTV cameras is much more. To eliminate this associate cost with CCTV cameras is the other solution is suggested such as GPS, Wi-Fi or Bluetooth, GSM, etc. This contact tracing is concerning positions. The position may be indoor as well as outdoor. For outdoor signal positioning, GPS provides a prominent solution. This is a more reliable technique to trace contacts. This provides accurate longitude and latitude. But it is only available for the outdoor systems only. In the case of an indoor system, it is not a suitable solution. To trace the contacts for indoor systems, the Wi-Fi or Bluetooth technology can be used. These use RSSI to check the distance among the connected devices.

The way of tracing contact is a more important term. Depended on the way of tracing the exact count and exact contacts can be traced out.

Suppose the population of N individual nodes comes in proximity for a particular time, then it is to be traced. For that network, the graph has to be drawn for the individual node. This graph will contain the connected nodes to the graph for a certain amount of time. From this, the degree of contact can be estimated at a specified time. This can be possible by using the following equation [30]:

$$k = \frac{1}{N}\sum_{i=1}^{N}\left(\frac{1}{T}\int_{o}^{T}k_{i}(t)dt\right)$$

where; N is an individual node; T is contact time; and k is the average degree of contact in a network over time T.

From that infection transmission can be calculated using the following equation [30]:

$$K_{i}^{*} = \sum_{j=1}^{N}G_{ij}(t)I_{j}(t)$$

where; Gij is the contact graph; and $Ij(t)$ is 1 if individual j at time t which can infect others.

10.4   PROPOSED METHODOLOGY

The proposed methodology uses Wi-Fi signals. These Wi-Fi signals are used to scan the smartphone connected in proximity. The social distancing application is installed on a smartphone. After installing this app when the smartphone will be carried by the person at the workplace, hospitals, etc. This will notify the proximity violation status. By considering that all people will be keeping their smartphones Wi-Fi hotspot TURNED ON. This will check the RSSI signals of connected devices in proximity.

Figure 10.4 shows the way of devices connected to the system. The central smartphone is connected with other devices through the Wi-Fi hotspots. It will be checking how many devices are connected to the hotspot and at which distance are they? If the distance among themselves is less than 2 meters, it will be sending a voice message to the Hotspot Turned on devices. If the rest of the devices are only connected through Wi-Fi, then these devices will receive an alerting message from the device to which it is connected.

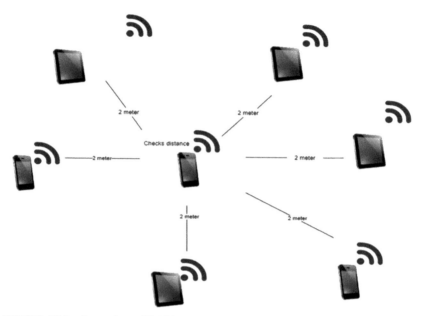

FIGURE 10.4   Smartphone Wi-Fi hotspot to alert the proximity using Wi-Fi connected devices.

This solution is possible by considering two ways such as:

- Personal social distancing app; and
- Workplace social distancing app.

The personal social distancing app is for the one who needs his/her security. This security is nothing but the notification to maintain the social distancing themselves. Also, maintain the log of connected proximity devices. A personal social distancing app needs to be TURN ON Wi-Fi Hotspot. By TURNING ON Wi-Fi hotspot will be itself informing to the user to which devices it is connected and which devices are violating the conditions. Whereas in case of the Workplace social distancing app needs the Wi-Fi hotspot turned on, which will indicate the connected devices and the Hotspot device will be notified to the other devices about the social distancing violating message.

10.4.1 THE PERSONAL SOCIAL DISTANCING APP

The personal social distancing app will be working as per Figure 10.5. In this, it calculated the distance among two devices by considering RSSI. This RSSI is used to the calculated distance among transmitters and receiving devices. The device which is present at the center will be acting as a transmitter and remain devices in radius distance they are acting as the receivers. This will continuously monitor the distancing among transmitters and connected different receivers. Once the receiver is trying to violate the rule, then it will be notified to the transmitter. The transmitter may also notify the violating node that maintains the social distancing and if the rule violated and the person comes into proximity either 1.5 meters or less than 1.5 meters then it will keep maintaining a log at Wi-Fi Hotspot TURNED ON devices.

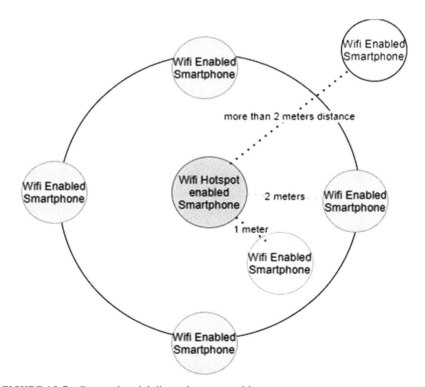

FIGURE 10.5 Personal social distancing app working.

Figure 10.6 shows the workflow of the proposed methodology for the personal social distancing app. The working flow of the personal social distancing app is explaining the alerting system mechanism.

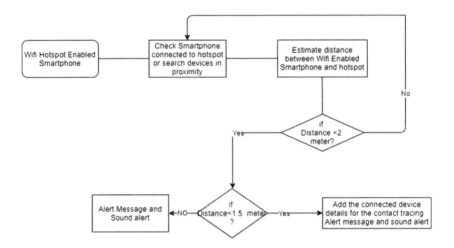

FIGURE 10.6   Personal social distancing app workflow.

10.4.2   WORKPLACE SOCIAL DISTANCING APP

The workplace social distancing app is working on the same principle. This workplace application needs some minor changes. Those changes are shown in Figure 10.7. In this, the Wi-Fi connected devices will be connected to the hotspot. This hotspot will continuously monitor the distance among different devices. After analyzing this distance amongst the distance, it will send an alert message to both connected devices to maintain the social distance.

This can be used in the workplace, at public transportation systems, and in schools also. Due to which contact tracing is easily implemented.

Wi-Fi connected devices that will checks devices connected in proximity. This app will be checking the devices which are violating the social distancing condition. It will alert by sending altering voice messages and also maintaining the log of connected devices in proximity. So that the contact tracing is easily possible.

Suppose any of the devices are still in proximity then it will note down the details of the connected persons. The distance between the two devices is calculated using the trilateration technique [25].

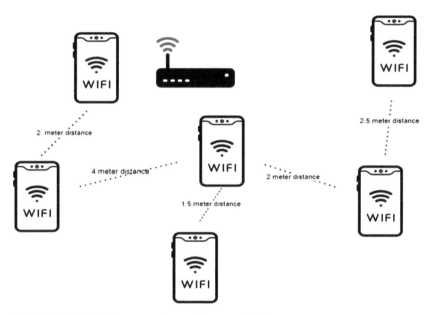

FIGURE 10.7   Wi-Fi connected smartphones to Wi-Fi router.

This contact tracing limits the spread of COVID-19 after analyzing the Network graph, and this graph will be generated for all the devices and stored for every node for 15 days minimum.

Due to this, tracing of the contact easily possible where the data is available at the authority. These authorities can implement the different social distancing policies by analyzing the logs and can reduce the contacts. This can be implemented in schools, colleges, hospitals, railways, the bus wherein minimum space maximum people occupy the space. Or we can say that the place where the transmission rate will be much more. The execution flow of the activity is shown in Figure 10.8.

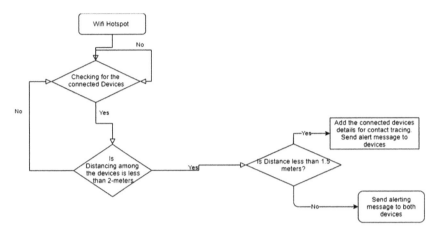

FIGURE 10.8   Working of Wi-Fi hotspot monitored by the workplace authority.

10.5   DISCUSSION

The key findings on the available data of the cases of infected and uninfected contact provided which indicates the isolation and contact tracing reduces the infectious in the community reduces the transmission rate. The impact of isolation and contact tracing is considerably dependent on the number of asymptomatic cases [31].

Due to this, the rate of transmission of infection among the people can be minimized. This transmission minimization is shown in Figure 10.9 and explained in Table 10.6. The comparative analysis in graphs shown in Figures 10.10 and 10.11. This transmission rate is considered only based on each person is maintaining physical distancing accurately besides with the usage of a mask covering the nose and mouth. The equation to minimize the rate of infection can be expressed by the following expression:

$$I = (R - x)^d$$

where; Ip is the number of infected people; R is the actual transmission rate; x is the number of people using the proximity barrier; and d is the number of days.

From this decrement in a rate of infection will be calculated by the following formula:

$$\%D = (Ip/I) * 100\%$$

where; Ip is the number of people infected after using the proximity barrier; I is the number without using proximity barrier; and D is the decrease in the percentage of the infection.

Figure 10.9, shows that transmission rate R0 = 2, On each day single person, is infecting only 2 people, and if suppose the among this two-person only one person is using this Wi-Fi-based proximity altering application then on day 1 only 1 person will be effected on next day again one person meets to two more people among then one is using Wi-Fi-based proximity app then it only affects 1 person. Hence the transmission rate lowers down from 2 to 1. This leads to a lesser amount of transmission.

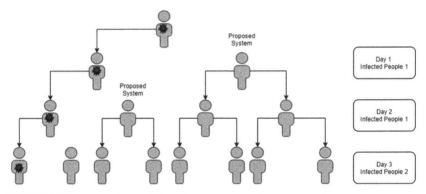

FIGURE 10.9   Assuming an infected person comes in contact with two persons among them, one person uses the social distancing app and follows the guidelines. Due to this R0 is automatically decremented.

Due to continuous alerts and messages, the probability of violation of safe distancing minimizes, and due to this, the transmission rate can lower down. This lower down in a transmission rate is shown in Table 10.6.

The speed of transmission falls below then only the speed of the spread of the virus could be limiting. This will increase the doubling rate due to which the transmission of the virus will take more time to infect the people. By using this barrier, asymptomatic infectious patients cannot act as the transmitter. Because asymptomatic patients are responsible for transmitting disease. This will limit the transmission because of which the doubling rate can be controlled [22].

The quick contact tracing, the contacts with the infectious person can be analyzed from the Network graph. This network graph stores each connected node for the stipulated time. This technique provides more accurate results. It helps to lower down the transmission rate. NCCU used the smartphone-based contact tracing due to which reproductive ratio after lockdown is between 0.01 and 0.04 which is relatively very small [30].

TABLE 10.6   Comparison between with Actual Transmission Rate without and with Social Distancing Barrier

	R0 = 2	The Barrier Used by 1 Person when R0 = 2	% Decrease in Infection	R0 = 3	The Barrier Used by 1 Person when R0 = 2	Decrease in Infection (%)
Day 1	2	1	50	3	2	66.67%
Day 2	4	1	25	9	4	44.44%
Day 3	8	1	12.5	27	8	29.63%
Day 4	16	1	6.25	81	16	19.75%
Day 5	32	1	3.125	729	32	4.38%

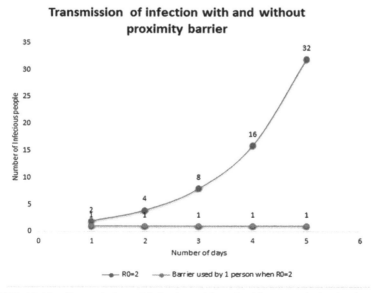

FIGURE 10.10   Transmission of infection with and without proximity barrier when transmission rate is 2.

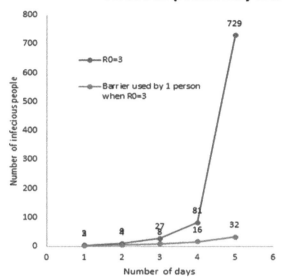

FIGURE 10.11   Transmission of infection with and without proximity barrier when transmission rate is 3.

10.6   CONCLUSION

To cope up with COVID-19 has the only solution is to maintain the social distancing. To maintain social distancing while doing daily chores is very difficult. Not only concerning keeping a safe distance but also to keep the track of a person who meets in the last 14 days. This is required for the Government to do contact tracing. Due to which contact tracing the transmission of COVID-19 infection transmission can be controlled up to cluster transmission.

Once the transmission mode changed from cluster to community to control the outbreak in densely populated countries like India will be more difficult. To this, the only solution is to have a proximity alert and contact tracing app. It covers the limitation of the Arogya Setu app which is keeping physical distancing among the people before infecting others. So, to minimize the transmission of the virus, this Wi-Fi proximity-based proximity alert will be acting as the light at the end of the tunnel.

KEYWORDS

- **artificial intelligence**
- **coronavirus disease 2019**
- **Ministry of electronics and information technology**
- **personal protective equipment**
- **severe acute respiratory syndrome coronavirus 2**
- **World Health Organization**

REFERENCES

1. Heath, K., (2011). The classical definition of a pandemic is not elusive. *Bulletin of the World Health Organization, 89,* 540, 541. doi: 10.2471/BLT.11.088815.
2. Worldometer, (2020). https://www.worldometers.info/coronavirus/transmission/ (accessed on 22 June 2021).
3. *European Center for Disease Prevention and Control: An Agency of the European Union.* https://www.ecdc.europa.eu/en/geographical-distribution-2019-ncov-cases (accessed on 22 June 2021).
4. *WHO Coronavirus Disease (COVID-19) Dashboard.* https://covid19.who.int/table (accessed on 22 June 2021).
5. European Center for Disease Prevention and Control: An Agency of the European Union *Transmission of COVID 19.* https://www.ecdc.europa.eu/en/covid-19/latest-evidence/transmission (accessed on 22 June 2021).
6. *WHO Coronavirus Disease (COVID-19) Coronavirus Disease (COVID-19): How is it Transmitted?* https://www.who.int/news-room/q-a-detail/q-a-how-is-covid-19-transmitted (accessed on 22 June 2021).
7. *Coronavirus Disease 2019 (COVID-19) Situation Report – 72* https://www.who.int/docs/default-source/coronaviruse/situation-reports/20200401-sitrep-72-covid-19.pdf (accessed on 22 June 2021).
8. *Unicef.Org,* (2021). https://www.unicef.org/media/82736/file/Considerations-for-school-related-public-health%20measures-in-COVID-19-2020.pdf (accessed on 14 July 2021).
9. Whitelaw, S., Mamas, M. A., Topol, E., & Van, S. H., (2020). Applications of digital technology in COVID-19 pandemic planning and response. *The Lancet. Digital Health, 2*(8), e435–e440. https://doi.org/10.1016/S2589-7500(20)30142-4.
10. Hammett, E., (2020) How long does Coronavirus survive on different surfaces? *BDJ Team, 7,* 14, 15. https://doi.org/10.1038/s41407-020-0313-1.
11. Cecilia, J. M., Juan-Carlos, C., Hernández-Orallo, E., Calafate, C. T., & Manzoni, P., (2020). Mobile crowdsensing approaches to address the COVID-19 pandemic in Spain. *IET Smart Cities, 2*(2), 58–63. doi: 10.1049/iet-smc.2020.0037 IET Digital

Library, https://digital-library.theiet.org/content/journals/10.1049/iet-smc.2020.0037 (accessed on 22 June 2021).

12. Hindustan Times, (2020). *Smart India Hackathon 2020: Defense Institute of Advanced Technology Wins First Prize.* https://www.hindustantimes.com/education/smart-india-hackathon-2020-defence-institute-of-advanced-technology-wins-first-prize/story-HL50sHUoAHPB1eeyG0ZHVL.html (accessed on 22 June 2021).

13. *How Different States in India are Using AI-Powered Tools to Combat COVID-19,* https://analyticsindiamag.com/how-states-in-india-are-using-ai-powered-tools-to-combat-covid-19/ (accessed on 22 June 2021).

14. *Observer Research Foundation Ideas. Forums Leadership. Impact.* https://www.orfonline.org/research/technology-governance-innovation-and-the-pandemic-69069/ (accessed on 22 June 2021).

15. ETGovernment.com. https://government.economictimes.indiatimes.com/news/technology/ai-powered-drishti-for-ensuring-social-distance/76638334 (accessed on 22 June 2021).

16. ETGovernment.com. https://government.economictimes.indiatimes.com/news/digital-india/covid-19-use-of-new-age-technologies-for-e-governance/75993744 (accessed on 22 June 2021).

17. Aarogya Setu: Government of India, How Does Aarogya Setu Work. https://aarogyasetu.gov.in/ (accessed on 22 June 2021).

18. mint, (2020). *Aarogya Setu App: How Bluetooth Helps in Identifying COVID-19 Suspects.* https://www.livemint.com/technology/tech-news/aarogya-setu-app-how-bluetooth-helps- in-identifying-covid-19-suspects-11587730877077.html (accessed on 22 June 2021).

19. Towards Data Science, (2020). *How to Build a Face Detection and Recognition System Computer Vision and Deep Neural Networks-Based Solution.* https://towardsdatascience.com/how-to- build-a-face-detection-and-recognition-system-f5c2cdfbeb8c (accessed on 22 June 2021).

20. Yi, L., Tao, L., & Jun, S., (2017). RSSI localization method for mine underground based on RSSI hybrid filtering algorithm. In: *2017 IEEE 9th International Conference on Communication Software and Networks (ICCSN)* (pp. 327–332). Guangzhou. doi: 10.1109/ICCSN.2017.8230129.

21. Kaibi, Z., Yangchuan, Z., & Subo, W., (2016). Research of RSSI indoor ranging algorithm based on gaussian - Kalman linear filtering. In: *2016 IEEE Advanced Information Management, Communicates, Electronic and Automation Control Conference (IMCEC)* (pp. 1628–1632). Xi'an. doi: 10.1109/IMCEC.2016.7867493.

22. Anderson, R. M., Heesterbeek, H., Klinkenberg, D., & Hollingsworth, T. D., (2020). How will country-based mitigation measures influence the course of the COVID-19 epidemic? *Lancet (London, England), 395*(10228), 931–934. https://doi.org/10.1016/S0140-6736(20)30567-5.

23. Centers for Disease Control and Prevention, (2020). *Contact Tracing for COVID-19.* https://www.cdc.gov/coronavirus/2019-ncov/php/contact-tracing/contact-tracing-plan/contact-tracing.html (accessed on 22 June 2021).

24. The Wire, (2020). *Will Bluetooth and Aarogya Setu Allow us to Safely Exit the COVID-19 Lockdown?* https://thewire.in/tech/bluetooth-aarogya-setu-covid-19 (accessed on 22 June 2021).

25. Rusli, M., Ali, M., Jamil, N., & Din, M. M., (2016). An improved indoor positioning algorithm based on RSSI-trilateration technique for internet of things (IoT). In: *2016 International Conference on Computer and Communication Engineering (ICCCE)* (pp. 72–77).
26. Waltenegus, D., & Christian, P., (2010). Fundamentals of wireless sensor networks: Theory and practice. *Wireless Communications and Mobile Computing*. John Wiley & Sons, ISBN: 9780470975688.
27. Bancroft, S., (1985). An algebraic solution of the GPS equations. In: *IEEE Transactions on Aerospace and Electronic Systems* (Vol. AES-21, no. 1, pp. 56–59). doi: 10.1109/TAES.1985.31053.
28. *Government of India Ministry of Electronics and Information Technology: Order Notification of the Aarogya Setu Data Access and Knowledge Sharing Protocol, 2020 in Light of the COVID-19 pandemic*. https://static.mygov.in/rest/s3fs-public/mygov_159051652451307401.pdf (accessed on 22 June 2021).
29. Aarogya Setu, (2020). *Government of India Aarogya Setu: Government of India.* https://aarogyasetu.gov.in/ (accessed on 22 June 2021).
30. Hernández-Orallo, E., Manzoni, P., Calafate, C. T., & Cano, J., (2020). Evaluating how smartphone contact tracing technology can reduce the spread of infectious diseases: The case of COVID-19. In: *IEEE Access* (Vol. 8, pp. 99083–99097). doi: 10.1109/ACCESS.2020.2998042.
31. Qifang, B., Yongsheng, W., Shujiang, M., Chenfei, Y., Xuan, Z., Zhen, Z., Xiaojian, L., et al., (2020). Epidemiology and transmission of COVID-19 in 391 cases and 1286 of their close contacts in Shenzhen, China: A retrospective cohort study. *The Lancet Infectious Diseases, 20*(8), 911–919. ISSN: 1473-3099. https://doi.org/10.1016/S1473-3099(20)30287-5. http://www.sciencedirect.com/science/article/pii/S1473309920302875 (accessed on 22 June 2021).
32. Shi, Y., Long, Y., Lu, F., Xu, Z., Xiao, X., & Shi, S., (2017). Indoor RSSI trilateral algorithm considering piecewise and space-scene. In: *2017 IEEE International Conference on Smart Cloud (SmartCloud)* (pp. 278–282). New York, NY. doi: 10.1109/SmartCloud.2017.52.
33. Cascella, M., Rajnik, M., Cuomo, A., et al., (2020). Features, evaluation, and treatment of coronavirus (COVID-19). In: *StatPearls [Internet].* Treasure Island (FL): StatPearls Publishing. Available from: https://www.ncbi.nlm.nih.gov/books/NBK554776 (accessed on 22 June 2021).
34. *The Size of SARS-CoV-2 Compared to Other Things*. https://www.news-medical.net/health/The-Size-of-SARS-CoV-2-Compared-to-Other-Things.aspx (accessed on 22 June 2021).
35. Pica, N., & Bouvier, N. M., (2012). Environmental factors affecting the transmission of respiratory viruses. *Curr. Opin. Virol., 2*(1), 90–95. doi: 10.1016/j.coviro.2011.12.003. Epub 2012 Jan 4. PMID: 22440971; PMCID: PMC3311988.
36. Ghinai, I., McPherson, T. D., Hunter, J. C., Kirking, H. L., Christiansen, D., Joshi, K., Rubin, R., et al., (2020). First known person-to-person transmission of severe acute respiratory syndrome coronavirus-2 (SARS-CoV-2) in the USA. *Lancet (London, England), 395*(10230), 1137–1144. https://doi.org/10.1016/S0140-6736(20)30607-3.
37. *Corona Kavach Is Government's New Location-Based COVID-19 Tracking App: This Is How You Use It.* https://gadgets.ndtv.com/apps/news/covid-19-

tracker-corona-kavach-   government-of-   India-coronavirus-how-to-use-2201703 (accessed on 22 June 2021).

38. *Centers for Disease Control and Prevention: COVID 19.* https://www.cdc.gov/ coronavirus/2019-ncov/daily-life-coping/contact-tracing.html (accessed on 22 June 2021).

39. Tang, J. W. et al., (2021). *COVID-19 has Redefined Airborne Transmission* (p. n913). BMJ; doi:10.1136/bmj.n913.

CHAPTER 11

Tracking, Modelling, and Understanding of Pandemic Outbreak with Artificial Intelligence and IoT

SAPNA KATARIA,[1] ANJALI CHAUDHARY,[2] and NEETA SHARMA[3]

[1]Assistant Professor-CSE, Noida International University, Plot-1, Sector-17A, Yamuna Expressway, Uttar Pradesh – 203201, India, E-mail: sapna.kataria@niu.edu.in

[2]Assistant Professor-CSE, Greater Noida Institute of Technology (GNIOT), Plot-7, Knowledge Park II, Greater Noida, Uttar Pradesh–201310, India, E-mail: anjali.chaudhary22@gmail.com

[3]Head of the Department-CSE, Noida International University, Plot-1, Sector-17A, Yamuna Expressway, Uttar Pradesh – 203201, India, E-mail: neeta.sharma@niu.edu.in

ABSTRACT

The growing need of the urban population gives birth to the concept of smart cities, which are equipped with smart devices and modern technologies for the collection and analysis of data. IoT devices collect data from sensors, electronics wearables, processors, vehicles, software, cell phones, etc., and examine using AI without substantial intervention from humans. These smart AI machines learn and enhance their performance with time-based on the data they analyze. Big data collected from IoT devices help in intellectual insights, to detect the patterns in the data using derived and dedicated algorithms. These smart devices help in integration of various government departments into a single platform that hosts different modern services efficiently and effectively. For example, traffic solutions, better health care, efficient parking automated systems, weather changes, and various other urban issues, including public safety. These smart devices can be extremely valuable in situations like the current outbreak of COVID-19.

This chapter is organized to explain the human-technology association in an efficient and effective way and offer practical explanations for monitoring virus spreads during pandemics. This chapter also discusses the use of Data Science in Pandemic Analysis and Management. This chapter further argues various ways for Pandemic Detection through Smart Computing Technologies and how to track the pandemic outbreak through Smart Sensor Networks. The Big-data collected here by various smart IoT devices can be used to Detect the Next Pandemic with the help of Smart Data Management. Smart data management here means World Wide Pandemic Data Analysis, Data Sharing, and Control through Smart Computing Networks.

11.1   INTRODUCTION

The world is facing irreversible damage of COVID-19 pandemic. Lakhs of lives has been already lost by the world, but the meter is running at a very fast at an uncontrollable speed. Apart from this, social, and economic damages have been there. The main issue in such a situation is that how can governments all over the world control the spread of virus or pandemic? Based on the research done over WHO reports, there can be two different approaches, techno-oriented approach, and human-oriented strategy to suppress the transmission of the corona-virus. It has been observed that even though the techno-oriented method possibly more adequate but it also leads to the repression of the civilian views. It is also highlighted that the technology with human interaction is facilitated and implemented by institutional and political framework.

Although techno-oriented method strives for universal acceptance of smart technology; a top-down approach is needed for implementing this technology into urban cities. Whereas, human-oriented technology requires governments to train their people and improve their communal assets, which will further help in developing and adopting smart technology. These technologies are the initial response of the government strategies to cope with the initial outbreak of pandemic. The human-oriented approach is still somehow inadequate and slower to control the situation whereas the techno-oriented approach was able to keep the situation under control relatively within the first three month of the outbreak according the reports published by WHO.

With all the smart technologies, AI algorithms can be developed to recognize the existence of new virus in humans but creating such algorithms is totally different from creating or developing other DL algorithms. The first stage of algorithm development, i.e., data collection become the main hurdle because Data sources are very limited at the early stage of any virus outbreak. Even after data collection, the next stage requires a huge amount of effort and expertise for cleaning and categorizing of the data as well as for ground-truthing of the data. The next main phase is training phase where collected dataset is used to train the developed models in which each data group is categorized by using large number of parameters and represented in a way such that training can be used to take a broad view of new cases in testing phase. The next phase is the testing phase where new dataset other than the training dataset; is used to test the outcomes of the developed model to regulate its success. These models are developed hypothetically leveraging the capability to adapt present AI models combined with clinical understanding to state new challenges of pandemic like COVID-19. The COVID-19 virus outbreak is a complex situation that involves human behavior, biological factors, biomedical companies, different governments, various government agendas, and its influence on healthcare, economy, social stability, and geopolitics. In the corona pandemic the data from hundreds of countries is analyzed everyday by the team of experts with help of a new AI analytical framework which will be useful for government in decision making about reopening of business and other economic and social activities.

To effectively manage the global pandemic situation like the COVID-19, expert needs to develop a Ranking system for countries based on the risk factor in that geographical area. This ranking algorithm can be developed according to the variety of medical facilities as well as non-medical factors. There could be some major factors including threat of infection, deaths due to virus, lifelong health conditions, hospitalization, negative economic factors of country, quality-of-life, etc. This analysis will be very useful to predict which countries have the highest possibility of positive outcomes throughout the global pandemic. In this type of analysis, a large number of parameters will be used further grouped into various categories

such as: virus spread risk, Government administration, Healthcare Adept-
ness, emergency treatment, disease management and Area Specific Risks,
etc., [3]. Some dynamic parameters should also be considered in these
algorithms according to the unique conditions of the geographical area,
which varies according to the countries such as highly consistent economy,
high level supply-chain, vivid tourism, efficient policies, etc.

The main concern of the experts in pandemic outbreak is the threat of
mistaken decisions by world leaders due to insufficient and incomplete
information. Incorrect assumptions could result in discriminatory activities
[5]. Urgency of the situation should not prohibit expert analysis predictions and honest calculations of uncertainty. Assumptions should not be
made on theoretical reasons, but statistical evidence of causalities should
be considered. For example, lower temperature data may be the result
of wealthier and better able to quarantined public, such data can lead to
analytical biases. Observational data must be evaluated cautiously because
expert findings would not be sufficient enough to rely on. In the early
stage of the COVID-19 pandemic outbreak, it has been already observed
that only accounting for people per square meter or considering the total
population changes the trajectory completely.

In human history, small, and distributed population has lethal and
local disease but globalization has turned viruses into mass killers. Over
100 years ago, SWINE FLU caused more deaths than world war. Modern
science must have cut down mortality rate, but various fatal diseases and
viruses like Ebola, influenza, bird flu, SARS, etc., are hunting mankind
over decades. Modern AI prediction algorithms can be turned into savior
by mining the data from social media and mobile phones. These optimized
search engine data can easily detect and track the Pandemic outbreak
through smart sensor network, where the disease outbreak and where it
may head next. Tracking mobile phone data has been proved a huge help
in other natural disasters also such as earthquake.

One more important thing about tracking a new virus or disease
outbreak is a diverse skill set. There must be a team for data analysis
including physicians, ecologists, veterinary, data scientists, geographers,
epidemiologists, biomedical persons, designers, and software developers,
etc. In recent, the rapid growth of the coronavirus globally has empha-
sized on the need of advanced big-data analytical tools in all sectors of the
healthcare industry. For comprehensive learning of the virus, researchers
are also emphasizing on the use of natural language processing (NLP)

along with artificial intelligence (AI) and machine learning (ML). This would certainly eradicate a lot of false trails and will allow identifying potential targets, instead of trying thousands of various things; it can be narrowed down to considerably smaller and faster which fast-track ultimate findings.

Another main concern of discussion in the COVID-19 situation is post-pandemic effects on social and economic structure. Are the effects being uneven or consistent for long term? To predict the after-effects of such a rare event, a dataset of a century should be included in research focusing on all major pandemics, armed conflicts resulting in major death poll. The most shattering pandemic of history "Black Death" has undivided attention of economists because of its crucial role in economic change all over Europe [2]. Data shows 25% to 40% labor scarcity, 5% to 8% drop-in land rates and significant rise in daily wages, but the question is how macroeconomics effects of Black Death represent pandemic like COVID-19.

11.2 HUMAN-TECHNOLOGY ASSOCIATION

Human-technology association is a term used to express the use of technology by human in different communities and societies. Human-techno association is especially observed in critical situations like pandemic COVID-19. World pandemic is the situation where the technology developed by human is being tested, i.e., how does human take advantage of technology to effectively and efficiently control the outbreak of the virus and to handle the massive burden on our societies. Within the past 100 years, we have already seen three fatal pandemics; COVID-19 is most recent with hazardous effects on communities and economies.

It has already been spread over the world with a mortality rate of 0.32% according to the data released by WHO. COVID-19 can be easily transmitted from human-to-human contact, and that invention of its vaccine would preferably take approximately 1 to 2 years. It has a substantial impact on approx. 80% of global GDP; more than 28.6 million populations have already been infected. So, WHO has recommended to the governments worldwide for quick action and vigorous surveillance to recognize infected cases, prompt isolation and 14 days quarantine. Inevitably, big cities are now hubs of rapid transmission of COVID-19 pandemic. Swift urbanization, population escalation, and more travel around the globe contributed to the transmission. Due to increased usage of smart technologies like the internet of things (IoT), AI, and big data; current cities management is more robust than before. These metropolitan cities were greatly affected by the COVID-19 pandemic in the first three months of its transmission.

All countries either implemented techno-oriented or human-oriented approach in their smart cities. While, techno-oriented approach strives for global adoption of smart technologies by the use of top-down method; which has been proved to be effective and efficient even at the first stage of implementation. On the other hand, human-oriented approaches strive for educating the residents and improve their communal and human assets for developing and adopting smart technologies. These technologies were employed by central governments as their preliminary response strategies. In the initial three months when the virus was recognized, countries with techno-oriented approach were able to control the first outbreak, but the countries who opted for human-oriented approach were continuously struggling in controlling the transmission. Considering these facts, current news articles and WHO reports reviewed the human-oriented approach as inadequate. Further research finds out that by using most sophisticated IoT and AI technologies and strict surveillance in the cities, the governments which adopted techno-oriented approach were able to retain the transmission under control, for example, the Chinese government. These researches claim that the effect of human-oriented approach is comparatively slower in transmission control like western countries [1]. It is also stated that while these new-age technologies are capable of enhancing the robustness of the government, human involvement will surely limit its potential as it is seen in Human-oriented approach.

As for technological perspective, smart technology as IoT collects data from sensors, electronics, mobile devices, wearables, vehicles, actuators, processors, software, etc., these devices collect information of the individual from different locations in the city and then this collected data is analyzed by AI technology, in simple terms, AI technology denotes the intelligent machines which learns from their processed patterns in the provided data, improve their performance and thinks independently without any interference from human [4]. For example, vehicles information, data from mobile phones, and information from cameras alongside the roads are used to update traffic situations and offer traffic solutions. Likewise, health care/hospitals records are evaluated to comprehend patients' circumstances and underlying health affairs to suggest better treatments. Big data has been recognized to be exceptionally valuable in the medical sector and throughout the outbreak of COVID-19. However, for the government, the extent of technology implementation is measured by the approach they adopt whether techno-oriented or human-oriented.

11.2.1 SMART TECHNOLOGIES USED IN TECHNO-ORIENTED APPROACH

Government's interest in adopting a techno-oriented approach is basically based on the extent of smart technologies already implemented in smart cities. To handle the growing urbanization, many governments across the world had already opted for many smart technologies, which proved to be a plus point for implementing techno-oriented approach. The best example for this is the Chinese government. According to an estimate, there are almost 500 world-class smart cities in China. Wuhan, one of the most affected cities and the center of virus outbreak is a smart city itself. Most of the political powers concentrated to the central government, the central government has control over all the provinces, which permits the central government to swiftly make preparations for coordination. At the grass-root level, IoT technology is so much penetrated into the daily lives of Chinese that commuter bus journeys are charged based on their facial recognition. Additionally, IoT devices are also linked to the utilities like dustbins, which automatically alerts authorities once they are full. Hence, both the environment and residents in major cities are focused on the implementation of IoT for almost 10 years now [1]. With new-age technologies like AI, IoT, and big data, city government can continuously monitor, make quick decisions, and offer disaster warnings. And for other sectors, the medical sector has also been profited considerably. Having a centralized political system, the Chinese central government quickly steps up to control the outbreak and transmission of coronavirus in Wuhan city. The IoT ecosystems were actively set into motion by the central government to bring various important stakeholders together. Different IoT devices were used to trace down and track out the infected cases. Following WHO's suggestions of identification, isolation, and lock down those areas which have more infected cases, the Chinese government use its deep-rooted surveillance structure and draconian technologies.

Maximum efforts of the Chinese government were put in identification of infected cases. A particular benefit Chinese government had in controlling the transmission was the surveillance structure developed by the government. The latest article stated that the maximum of the investment in cities was invested in building surveillance structure on the citizens, having almost 160 cameras over 1000 residents. Chinese government collected data from the sensors all over the cities; conducts distributed testing, and flag infected areas. The development of the new robust AI technology to identify infected individuals helps the government to complete the job of 15 minutes in only 10 seconds. Drones with cameras were deployed for regulating surveillance of citizens and for issuing guidelines and admonitions to those citizens who fail to follow government instructions in the emergency period. Government strategies also included installing IoT devices to collect travel information of citizens in infected areas; the travel history of people to the corona affected cities was supervised and communicated to the concerned specialists via AI technologies. Public transport system was also installed with precise technologies for identification of those passengers who are expected to be infected.

New AI devices were developed to scan the body temperature of passengers because very high or very low body temperatures are symptoms of COVID-19 [1]. A new AI screening device scans almost 15 patients in just one second; a range of these devices is a maximum of 3 m developed on the concept of on contactless remote screening system. Another same technology device developed for scanning body temperature that can perform screening of almost 200 people each minute. These contactless AI devices were installed all over cities in public places for temperature detection.

Moreover, police personals in smart cities wear new techno-helmets which can identify anyone with fever up to a 5-meter radius. Largely, the IoT technology and AI systems were extremely competent and a cause of comfort for the overburdened health organization and government in China.

11.2.2   SMART TECHNOLOGIES USED IN HUMAN-ORIENTED APPROACH

Unlike China, western democracies have opted for human-oriented approach to tackle with COVID-19. In western democracies, political powers are shared between regional and central government. Similar to China, western democracies also have well connected smart cities which have great trading connection all over world including China. Trade is the main reason for the spread of COVID-19 from China to western democracies. The human-oriented approach is concerned with citizens; in this approach, technologies are particularly used. This approach allows citizens to relish their freedom. This approach can also be termed as 'soft technological determinism' in which people hold intervention or assistance in deciding technological use. In this approach, governments also improve societal and human assets among their residents, which permit the development of new technologies as per local needs. The human-oriented approach permits governments to implement new-age technologies to guarantee better residents' engagement, to propose responsibility and modernize connections between societies and government. Moreover, residents seek better involvement in the preparation of urban growth strategies, frequently beginning local enterprises. The difference in techno-oriented and human-oriented approaches possibly embraces a key for deciding the level of new-age smart technologies used in the framework of both Western democracies and the Chinese government to control the transmission.

In contrast with human oriented approach, the techno-oriented approach mostly focuses on the significance of smart technology and makes residents subordinate to technologies. Techno-oriented approach expects technologies to fix all malfunctions in cities. This approach can also be termed as 'hard technological determinism' that considers technologies as the root of society. The essential abilities of smart technology are to solve complications experienced by governments. While Techno-oriented approach is more effective in striking law and order but it also restricts human freedom, heighten suppression, and it raised moral questions. As a protection against COVID-19 spread, WHO initiated certain guidelines like washing hands regularly, using alcohol-based sanitizer whenever soap and water is not available, while outdoor avoid touching your eyes, nose, and mouth, follow social distancing, avoid traveling unnecessary, avoid social gathering, wear mask when outdoor and suggested to try to enhance their immunity. All the government which opted for human-oriented approach requested their citizens to follow WHO guidelines for example, Indian Government. These governments work intensely at both state and central level to minimize the cases. In a human-oriented approach also governments are using new-age technologies AI and IoT, Big data and data science, collecting, and analyzing data, providing modern medical facilities. The only difference is that in these countries' government did not try to suppress the fundamental rights of citizens. In this rapidly growing deceptive pandemic war government is fighting hard with citizens with the help of medical consortium, NGOs, police, and paramilitary forces. Although with the increasing number of cases daily, there is a persistent requirement to control the COVID-19 combat at a technologically advanced level. Now, the Indian government is getting extremely involved with the scanning and isolating every infected person. At present, the testing technology facilitates immediate PCR test, Point-of-Care fragment analysis, prompt antibody test which is appropriate for observation as results take 7–10 days after infection and Point-of-Care quick antigen identification test for initial recognition of COVID-19. Indian Government has developed 1105 new operative labs in order to take more number of tests every day. Furthermore, more than 2 crore N-95 masks and around 2 crore PPE kits have already been distributed Indian government for free till now. Indian government launched 'Arogya-Setu' mobile app on which citizens voluntarily share their information with the government for tracking. Developing countries are still struggling on how to handle COVID-19 while taking care of their economy; they cannot afford to lock down the whole country or major cities, which will surely increase hunger and unemployment issues and adversely affect their economy.

Table 11.1 shows the differences between both approaches opted by different governments all over the world.

WHO recommended and opted for three hypotheses-identification, isolation, and separation for decreasing the transmission, and seven other hypotheses and smart technologies usage to explore techno-oriented or human-oriented approaches. The hypotheses that incorporate governance are smart technologies like IoT and AI, confidentiality concerns, lockdown of countries or cities, activism, sharing of relevant data, and epidemic of information [4]. These hypotheses are concluded from the shared information, to understand whether government will opt for a techno-oriented or human-oriented approach.

TABLE 11.1 Techno-Oriented vs. Human-Oriented Approach

Parameters	Techno Oriented Approach	Human Oriented Approach
Effects	Effectively and efficiently controlled the situation	Lack of coordination caused increase in number of cases
Government	Government relies on technology	Government relies on their people
Privacy	Individual's privacy compromised	Nobody's privacy is compromised
Quarantine	Very strict lockdown imposed with the help of IoT devices and AI technology	Limited use of technology plus manual lockdown
Surveillance	Individuals' mobile devices were tracked whether they are infected or not	Government relies on service providers for data-keeping individual's privacy safe
Fundamental rights	Fundamental rights were compromised	Fundamental rights of public were never compromised
Sharing information	People were forced to share their data	People volunteer for sharing their data otherwise no-one was forced to share any information
Strategy	Everyone was under house arrest	Partial lockdown was adopted
Government response	Initially denied or suppress the information	Initially denied but communicated to public
Coordination	Well-coordinated by central government	Lack of coordination between central and state government
Activism	Activists were suppressed and under home arrest	Activist were free to question government about their rights

11.3 DATA SCIENCE IN PANDEMIC ANALYSIS AND MANAGEMENT

Data Science is a technique about data-mining, data preparation, exploration, imagining, and preservation of data. It is just not about computer science, but it is an interdisciplinary area that uses technical methods to process and extract perceptions from data. With computer science technologies it also requires knowledge of mathematical and statistical tools and domain knowledge of various fields so that various regression and classification models can be applied for better research and analysis of that field. It involves extraordinary capabilities of a 'data scientist' for using various ML methods to apprehend and evaluate data. The data scientist not only evaluates data but also make prediction for future incident with the help of ML algorithms. Nowadays data has turn out to be the new fuel of engineering and industries. It has become the new electric current for AI and IoT. Each and every area engineers and industries need data to operate, propagate, and upgrade their businesses. The data-oriented approaches adopted by industries originate meaningful perceptions with the use of data science.

These perceptions would be very useful for the industries who want to evaluate pandemic information for developing new technologies and their performance. Healthcare engineering also use Data Science Technology in abundance to diagnose nanoscopic tumors, abnormalities, viruses, and pandemic outbreak at the initial stage. The primary usage of data science in analyzing pandemic data and managing medical industry is by various medical imaging techniques like X-Ray, MRI, and CT scan. Regardless of continuous efforts to advance medical systems universally, developing virus pandemics persist to be the most important public health apprehension. Effective and efficient response to these virus outbreaks depends on appropriate intervention on time, preferably informed by each reliable data source. The collection of reliable data, analysis, and visualization of data is pretty complex. With increasing data every day from various unreliable sources, the complexity of generating interesting patterns from that data is also increasing. An evolving data science is focusing on the technical and operational phases of the transmission data pipeline, starting from collecting to analyzing and reporting information of the outbreak. The main concern should be the analytics modules, their dependencies on each other, data requirements; analyzing the outbreak responses and generation of interesting patterns which can provide enough information to notify efficient procedures in existent time. In this chapter, we will learn the perspective of virus transmission based on the current situation with the help of data science. The data science analyzing these information sources involves a comprehensive series of approaches, together with database schema design and portable technology, frequency statistics and max probability assessment, collaborative data visualization, bio-statistics, Bayesian statistics, graph theory, mathematical modeling of information, genetic exploration and evidence fusion approaches [6].

11.3.1 COLLECTION OF DATA

Pandemic data has to be pre-processed, selected, and sorted into two categories "infected case" and "general information." Data of infected cases comprises of description of patients gathered where every row of table holds infected cases and every column holds base-line information like age, gender, location, travel history, family member's information, etc., thereby satisfying the description of "uncluttered data" in data science society [7]. "General-data" documents the unique attributes of the infected population. This document Contains population statistics information like population density, age hierarchy, mixing patterns, traveling history data, healthcare amenities, medicine stocks and epidemic data, pre-existing immunity information. One extra data file which we should consider is 'interference data,' which maintains the information about government decisions and efforts arranged for interference and post-interference like vaccination, treatment provided, isolation, active cases, and other actions [6].

"Infected-cases" data can be gathered into three categories, individual information category, exposure to others category, community spread or clusters of cases category [8]. In the first category, every infected individual's personal data will be collected like any major medical issue, breastfeeding or pregnancy, recently operated, etc. In the second category, data about other persons is collected who may have exposed to the infected person like family members, colleagues, relative friends, etc. In the third category, collective data from the societies or cities is taken into consideration to further evaluate the reason of increasing cases. In some countries, the only reason is community spread, i.e., person to person transmission, while in some countries, clusters of infected persons are found in different areas or cities. Even though in these categories some data is environment or conditions specific, others data may be specific to the resident's proportion. This data will be beneficial for extensive range of pandemics which can be consistently and centrally gathered in analyzing the current situation and be prepared for future outbreak.

11.3.2 PANDEMIC ANALYSIS TOOLBOX

For analysis of pandemic data, various methods and tools can be used to gather and analyze information, visualize, and model that information, and to generate reports on gathered data. These tools and inter-dependencies between them are précised in a typical workflow derived from scanning pipelines used throughout the latest pandemics. The workflows can differ considerably in other pandemic situations for example, food-borne virus transmission may concentrate on backtracking of data, whereas vector-borne virus outbreak may focus on demonstrating the vector's biological function. Latest technological tools include mobile data gathering tools, cloud computing, electronic information collection and automatic data evaluation and report generation. These tools widely use available low-cost hardware like mobile phones and tablets; they consume less power and gather data offline, thus making these devices a great resource pool. These new electronic devices can also increase data excellence by the practice of control rules and reasonable checks, and by regular reporting even if there is no case. These electronic devices can decline the gap between information collection, centralization of data and evaluation, which is crucial for data-oriented feedbacks. In addition, electronic devices facilitate the gathering of additional information including GPS location coordinates, barcodes, and pictures beneficial in linking case info with clinical samples and in assisting diagnosis by openly interfering with analytical devices. These devices also save our time with the help of 'form logic' like automatically skipping parts of the survey or study which are not significant to a particular participant, whereas real-time, automatic centralization, data exploration, and generating reports can be built directly into the electronic platform. Electronic device systems offer chances where unauthorized interference and exposure to the personal information of people, although many electronic systems provide end-to-end cipher code technique throughout data transfer [9], but very few provide added security by encryption at data entry level.

11.3.3   DESCRIPTIVE EVALUATION OF INFORMATION

The most essential phases in data evaluation are exploration, whereas visualization has a central part, which is accomplished with instructive summary data [6]. The graphics are required for quick calculation of currently running dynamics, i.e., epidemic curve, which demonstrate case frequency as histogram for a certain time period [8]. Growing case tallies should not be included in epicurves because they have a tendency to make current dynamics dubious and generate statistical reliance in data that may generate unfair results, which will obviously lead to wrong estimations. Nowadays, maps are normally used for visualizing the distribution of infected cases, for demonstrating the 'ecological statuses of infected areas at a huge scale. Mapping tools which can find out the areas of spread of virus include the geospatial positions of virus transmission, which also maps hazards of spread of virus [10]. It also focuses on recording virus evolution, pandemic transmission, and quick recognition of outbreaks. Concise or summary data is useful for data visualization in the data analysis phase. Some criteria like communicable need the use of mathematical models for estimation, and they are not freely accessible as graphic tools. Other useful information can be easily computed from descriptive data, including population statistics pointers of the infected cases like age, gender, occupation, etc., fatality ratios or case suspensions like hospitalization time, recovery progress, death, percentage of infection reported. Virus incubation period 'contamination to symptom' is the additional significant interval for notifying the intervention (like defining the period of contact tracking, declaring outbreak end), but then again, it is difficult to derive. Note that these are greatly analyzed by symbolizing the complete distribution such as probability distribution rather than stated as solitary central value [7, 8].

A very good estimation of communicability is the progress rate (say P), that can be easily calculated with the help of a simple workflow model in which frequency rate is either exponentially increasing (P, 0) or decreasing (P, 0). Normally, the range is evaluated straight from curves of a linear model; here progress rate is stated as the curve of the linear regression model [11]. Moreover, this approach is simple and computationally efficient, it also has the advantage of the linear modeling structure; thus, permit to estimate the uncertainty related with the given data. The log-linear estimation model can only be good with exponentially increasing or decreasing data of outbreaks that is not always suitable for real-time complex data structure. While the progress rate estimates the virus transmission speed, but it does not comprise of the intervention data amount required in controlling virus transmission. This should be better stated as the 'imitation number' (denoted as R), which calculate the average number of subsidiary cases produced by every primary case. On the other hand, Simulation estimation models include a detailed representation of all the various factors that may stimulus transmission. These models are a better choice for evaluating the predictable results of possible treatments, but they typically require cautious selection of parameters and rigorous computation [11], both of these functions are challenging at an early stage of the outbreak.

11.4 TRACK THE PANDEMIC OUTBREAK THROUGH SMART SENSOR NETWORKS

Smart technologies based on Sensor Networks signify the subsequent development phase in engineering, like industrial automation, traffic monitoring, video surveillance, and robot controller. Data from these sensors come from several networks of interrelated sensors with remote distributed geo-locations. Healthcare technologies are intensely developed as numerous modern treatments are available now, modern therapy techniques and cures, latest medical device, and appliance. Modern technologies are seeking for techniques that can improve the capability of the healthcare system, which can reduce human faults, which can simplify the treatment procedure, and which can improve the living standards for doctors and nurses. Wireless sensor medical devices can easily record the temperature, patient medical status, psychological data, saline level, heart rate, blood pressure, etc., and save all the data in real-time. Wireless technologies have become essential in our life, we can effortlessly detect wireless signal anywhere around us. Recently, wireless LAN, radio frequency, infrared (IR) waves, and Bluetooth signals are the most used and easily available wireless technologies which are usually used for transmitting data and receiving information as signals [12]. Every industry and area are widely using these because these devices can improve the working capabilities and decrease the cost of installing wired networks. The ZigBee is the new-age wireless automated telecommunication method that can transmit in different patterns, for example, point-to-point, point-to-multipoint, and peer-to-peer with a speed of 250 kbps [13]. According to ZigBee features and capabilities, it can be very useful to develop the wireless telecommunication sensor networks for industrial application software.

After testing ZigBee for heavy Industrial applications, it is recommended to use the technology for temperature monitoring of the citizens which is also very useful for pandemic control. This technology is designed to focus on nonstop temperature checking and the data communication. The execution of multipoint and peer-to-peer network could be established easily by organizing each and every module to function as a sensing node. ZigBee has also been studied with wireless communication devices such as Wi-Fi, Bluetooth, and microwave ovens. They have proposed a sensor network based on ZigBee technology for healthcare wireless application. It is the architecture of modern patient room with wireless sensors for monitoring the patient status and body temperature, level of saline, room free or not, nurse/doctor around or not, etc. Moreover, the working status of nurse, doctor visits, and patient movements are recorded and communicated wirelessly to admin for future use in situations like pandemic [12, 13].

Wireless devices with simple processors act as sensing nodes; consist of various low-cost detectors with less power consumption antennas. As these devices do not need wired infrastructure, they can easily form a network without much hard work and time. These sensing nodes can easily get in touch with each other to collaborate and perform complex jobs. These devices comprise of a sensing node, processor unit, transceiver, detection unit and a power unit with long battery life, which make these devices an ideal resource for real time data collection, data transmission, evaluation, and research purposes. The sensing devices collect uninterrupted data as analog signals, which are perceived by detectors and converted into digital signals with the help of "Analog-to-digital" converter and trans-mitted to the admin for further processing. These sensing devices should be capable of working with extraordinary physical intensities, should be working without surveillance, autonomous, and should be adaptable to the environment. Other than medical applications these sensing device were also experimented with the areas where wired networks could not be installed such as habitat monitoring for animals and plants both, weather forecasting, environment monitoring, power supply system monitoring, detecting movements and environment changes in remote locations like forest fires and enemy movements or terrorist attacks [14]. For wireless communication three most used technologies for pandemic data collection are Wi-Fi, Bluetooth, IR, and ZigBee. The positive and negative points of these technologies are compared, and it was found that although Wi-Fi provides many remarkable features but ZigBee is the most appropriate technology among these; because ZigBee fulfills the low-cost and low maintenance requirements of today's growing needs. In current situation like COVID-19 pandemic low-cost resources are most required for data collection and evaluation, to tackle current pandemic situation efficiently and to detect next pandemic at early stage; data is the most important feature which should be focused.

Table 11.2 shows the main difference of wireless technologies.

TABLE 11.2   Comparison of Wireless Technologies

Features	Wi-Fi	Bluetooth	Infrared	ZigBee
Bandwidth	Up to 50 Mbps	Up to 2 Mbps	Up to 120 kbps	Up to 250 Kbps
Transmission	Data, voice, and video communication	Data and voice communication	Data communication	Data and voice communication
Connected devices	Up to 250 devices can be connected	Many devices connected by personal network	Only two devices are connected	Up to 256 devices can be connected
Range	Up to 100–150 meters	Up to 10–20 meters	Up to 10–20 millimeters	Up to 100 meters
Stronghold	High data rate	Inter-operability	Line of transmission	Low-cost and maintenance
Network size	32	7	2	64,000
Applications	WWW, e-mail, voice call, video conferencing	Data transfer and voice communication	Direct and slow data transfer to another device	Monitoring and control
Battery life	Up to 5 days	Up to 7 days	Up to 10 days	Up to 100 days

Except ZigBee, the second most used technology in times of pandemic is Wi-Fi based on the features, strengths, and weaknesses. Wi-Fi also provides flexibility of choosing any topology for developing wireless networks; in the current situation, mesh topology could be proved a better option. Employing wireless communication techniques and sensing devices in smart cities for surveillance and predicting future events by evaluating the collected data will make us capable of detecting the next pandemic as well as providing enough information for handling the current pandemic situation. Researchers and scientists are gradually increasing the use of big data, ML, AI, and NLP to track coronavirus (CoVs), and to gain a further inclusive understanding of this virus. From starting months when COVID-19 start to spread out of China, researchers are working very hard trying to discover the physiology of coronavirus. Some questions are really need to be answered like why this virus has different effects on some people than others, what measures should be followed to decrease the spread, and what level this disease is likely to go next, what vaccine can cure or control the effects of coronavirus. All this research is based on the big data collected directly from corona patients.

Table 11.3 shows the strength and weaknesses of wireless communication technologies used during a pandemic.

TABLE 11.3 Strengt hs and Weaknesses of Wireless Technologies

Technique	Strength	Weakness
Wi-Fi	• Easy implementation	• Comparatively low security
	• Low-cost and convenience	• Only useful in customer local range
	• Mobility and expandability	
	• More productive than wired	• Low-Reliability and bandwidth
	• Permits higher data rates than ZigBee	• Most devices do not allow Bluetooth connection
	• Many products available	• Low security
	• More mature technology	• Maximum number of devices is limited
ZigBee	• Low maintenance	• Limited range
	• Low implementation cost	• Comparatively low data rates
	• Better scalability	• Could be more secure
	• Maximum number of devices allowed is relatively large	
Infrared	• Point-to-point and secured communication	• Frequency affected by hard objects
	• Provide good stability over time	• Harmful waves
		• Lower data rate transmission
	• No corrosion or oxidation can affect the accuracy of infrared sensors	• Only two devices can be connected

Positively, in a recent survey a group of data scientist and researchers regarded medical field as the region where data analytics could have greatest potential, 73% believed that big data analytics could definitely avoid future pandemics. Another promising approach is data mining, whereas AI can eradicate many deceitful tracks and permit us to recognize promising targets. This approach may prove an excellent option for researchers; instead of scanning 1000 different records, it can be reduced to a miniature size much faster. That is eventually going to speed up the discovery of the corona vaccine. Although data analytics techniques are the best option to any progress in pandemic analysis but experts do not always get the right information; to really succeed in this pandemic situation, experts from multiple disciplines need to bring together who have the knowledge of AI, computation, and ML and those who can apprehend the biomedical and biological implications.

11.5   DETECTION OF NEXT PANDEMIC BY USING IOT DEVICES

From December 2019 till date whole world is fighting a serious pandemic called COVID-19. It is not affected our health (mentally and physically) by also effected out but economic. These pandemics create gigantic challenges for public health officials and government to collect information as quick as possible and respond accordingly. As we are still suffering from this pandemic COVID-19, there is a need to be surveillance, prevention, and rapid-response efforts to slow down and stalling outbreak and be alert and prepare for any new pandemic situation if arise. The Centers established for Disease or virus Control requires to be observed in most concerned areas. Substantial efforts are invested in the improvement of cost-effective, user-friendly, and portable systems for diagnostics at a rapid rate, where we can implement an IoT for healthcare via a worldwide network. Conversely, IoT based instrument is not available at present scenario.

In such a situation, IoT, AI (Artificial Intelligent) and ML can play a vital role for predicting and minimizing the pandemic outbreak. Today, the enormous world population using the smart mobile phones connected to mobile networks via 2G, 3G, and 4G which are embedded with GPS (global positioning system). By using mobile network, maybe we can create and implement a cost-effective by using the technology Internet of Thing. The DENV analyzes would create an IoT via a mobile network, for medical facilities by joining the "easy-to-use" and "cost-effective" (point-of-care) POC TOOLS, which is a crucial tool for managing any virus outbreak [16]. By using big data analysis of different models, Centralized location can be used to store the information from POC devices by using IoT to speed up the information transfer and suggests the affecting sites. As suggested, the IoT devices are capable of communicating with the Point-of-Care Systems. It can inform its machinist where to go next to perform the test. Last year 5th generation of network was released with an enormously increased data rate compared to previous generations like 4G, 3G, and 2G. Even though, DENV testing can rely on only IoT devices for POC applications because it required a low data rate. A method has been proposed for the recognition of virus outbreaks at an early stage with the help of Internet. This approach delivers a proficient intelligence method for data-collection and offers productive solutions, but then the need for high-speed steady internet access and inadequate evidences of identifying threats; it can be regarded as the weak point of this proposed method [17]. Another method has been investigated for recognizing the spread of flu with the help of social-sensors in China. They make a strategy for identifying the communication of virus built-on social networks [18]. The major weakness of this proposed method is the high-resource prerequisite for information processing and prediction.

IoT have been identified as the main key for error reduction in detecting and handling communicable diseases. The records of patients from various geographical places can be efficiently aggregated and accessed with the help of innovative technology [19]. The proposed solution is based on IoT combined with RFID technology for appropriate identification. The major down-side of this proposed cloud is the absence of security, non-standardization of communiqué-protocols and rising of scalability disputes. A smart IoT system was recommended for virus monitoring that affords efficient and effective monitoring, detecting, and tracking the virus trends. On the other hand, it has some inadequacies including failure to integrate with another device, high-execution cost, and insufficient feed-back mechanism. It also involves improving the use of devices for timely data-collection and accurate prediction of pandemic [20]. A more secure cloud framework based on IoT was proposed for supervising Zika virus-outbreak. This proposed framework works efficiently for discovering Zika-virus but absence of generalizable in detecting future epidemic outbreaks [21]. As author discussed, if we can update the frame as suggested can work for pandemic like COVID-19. Successful execution of predictive modeling can be representing a key leap forward to fight and get rid of the world from some of the significant infectious diseases. Big-data analytical can decentralize the procedure and facilitate the timely investigation of wide-spread data produced by IoT and mobile devices in real-world [15]. Some other suggestions are also given below:

11.5.1   BUILDING INTELLIGENCE AND KNOWLEDGE

Big-data and artificial-intelligence analytics have major roles in the latest genome-sequencing procedures. High perseverance computer-produced simulation allows the researchers to learn and understand huge datasets about virus study and their spread. A big consideration of these spectacles can be empowering the public to react far more speedily to the attacks. Even though IoT has generated unparalleled prospects that can upsurge revenue, decrease costs, and upgraded efficiency, gathering a huge amount of data alone with IoT is not enough. To create remunerations from IoT, a platform must be created where data can be easily collected, managed, and analyzed an enormous amount of sensor data in an expandable, accessible, and affordable manner [30]. In this perspective, purchasing a Big-data platform that can easily merge and consume different data from various sources as well as in hastening the data-integration procedure turn out to be vivacious. An IoT big data-analytics structure has been proposed to overwhelm the challenging tasks of storing and analyzing huge data generating from smart buildings. This framework comprises of three modules which are IoT sensors, big-data management, and data analytics. These analytics are accomplished in real-time to be helpful in numerous departments of smart-building management like managing oxygen levels, hazardous gases, smoke, and brightness. This framework was implemented at a platform "Hadoop-distribution" and the analytical were performed with the help of "PySpark." The results of this platform show that the provided framework can be useful for IoT-enabled big-data analytics [31].

11.5.2   AUGMENTING MEDICAL CARE

In recent times, heartbreaking images of medical personals were seen worldwide working vigorously to save COVID-19 patients, regularly laying their own lives at risk. In this scenario, AI, and ML can play a crucial role in the reducing their burden while guaranteeing that the good quality-life should be there for health workers and better treatment options for patients. Several most favorable usages in medical region in AI include precision medicine, predictive analytics, clinical decision support and image-diagnosis. These areas are still under development requiring huge investments in AI, which have increased over some past years [1]. At present, medical-AI systems have already started to offer great value in the dominion of deep learning, pattern-recognition, and neuro-linguistic programming (NLP). ML systems were developed to detect errors in data without propagating them. Nevertheless, medical-AI systems could never substitute the real-physician need or the relationship between them. These systems enhance medical care provided by real physicians and do not substitute it. Medical-AI systems should be reproducible, transparent, and trustworthy for both medical professionals and patients. However, these systems must emphasize the needs of the users, and Usability must be verified by applicants who replicate comparable needs and exercise trends of the final user, and these frameworks must perform excellently with societies or individuals. Physicians are more likely to agree with an AI system which can integrate or mend the current exercise patterns and patient care. As author given an example of Tampa General Hospital located in Florida, they are using AI to detect body temperature (fever) of visitors by a simple facial scan. Another example belongs to Sheba Medical Center located in Israel is also using AI to predict the complications like respiratory failure or sepsis during COVID-19 patients.

11.5.3   CHALLENGES

Inventors and supervisors of medical-AI systems should and must safeguard suitable disclosure and they must note the limitations, benefits, and possibilities of proper usage of these systems. Consecutively, physicians and healthcare professionals must understand medical-AI techniques and methods to rely on clinical recommendations. Training in the prospects of medical-AI systems must occur in presence of both medical-students and general-physicians; involvement of healthcare professionals is critical for effective evolution of these systems. Frameworks and systems based on AI must always follow ethics, principles, and integrities of the healthcare profession. Protecting the personal data of patients, confidentiality, control of data is an essential trait of patient and physician relation. Anonymizing the data does not necessarily be responsible for protecting patient's personal data when ML procedures can easily detect any individual from huge complicated data files with only a few data facts provided, that will surely put patient's privacy in jeopardy. Patient's expectations for the security of their private information should be considered critical in these modern systems that consist of patient's consent and data supervision by developers. Feasible solutions to alleviate the security risks have been explored and proven to be influential adoption for medical-AI systems. Structure of data and reliability are the most important challenges required to be considered while designing medical-AI frameworks.

Training datasets for machine-learning algorithms are designed by humans that could be bias or may have errors. So, the training data sets need to be normalized and error-free. Which in turn indicates that minorities might be a disadvantage because the data accessible about them is way more less for algorithm design consideration. Another issue with design-consideration model-evaluation methods for precision and includes very attentive analysis of training dataset and its relation with testing dataset is utilized for evaluating the algorithms. Accountability issues show substantial challenges when it comes to the adoption of modern AI systems in the medical field. As supervisors of existing systems and developers of modern medical-AI systems, typically they have the most prominent knowledge about security risks and are the best options to be selected for mitigating the threats. Therefore, medical professional associated with medical-AI systems must be responsible for unfavorable incidents occurred from technical malfunction. Physicians are generally found irritated with the compulsion of automated health records; because these systems are developed to support workflow patterns and team-based activities but often disappoint. Noteworthy consideration should be given to proper system implementations; it is not easy to deploy all the systems to every single setting as a result of variations in data sources. Development work of such systems is already ongoing to promote governance and medical-AI systems, involving principles for healthcare, certification measures, government guidelines, cognitive property rights, and legal and ethical considerations.

11.6   POST PANDEMIC EFFECTS

The after effects of pandemic on the social and economic life of people worldwide were always so worse that they cannot be explained. People across the world had suffered a lot; they have gone through the situations which could not be imagined. Some facts and figures have been shared here which cast light on the life of people after pandemics.

11.6.1   SOCIAL EFFECTS

Every time pandemic affects the social life of the globe in various ways. There is no area where pandemic has not affected the human social life including way of communication, transportation, business, education, medical, and many more. Social distance is the best prevention for the pandemics. So, people movement was restricted by implementing lockdown. People were not allowed to visit their relatives; they are allowed to go for market only for the essential things, for example, daily livelihoods and medicines. Airports, bus stops, railway station and borders were sealed so that so that spreading of pandemic can be stopped from one geographical location to another. Schools and colleges had been closed for months. Due to lockdown, industries, airports, shops, and other income sources of people were closed for months which affected people physically and mentally, and financially. Even many people lost their jobs. Old age people and persons with disabilities are most vulnerable to the diseases and do not have enough strong immune system to fight the disease. Moreover, old age people if they have any choric disease for example high blood pressure, diabetes they need extra care and precautions. Old age and disable persons not only deal with their health condition even it becomes so difficult for them to stay in isolation when they need extra care and support.

In the year 2020 whole world is dealing with COVID-19 or coronavirus. COVID-19 is an infectious disease which spreads from a person to another. Social distance is the best prevention for this disease. So, to maintain social distancing people adopted digital technology. Digitization is across the globe. Education and academics field evolved completely. This pandemic replaced board and chalk with laptop. Virtual classes were the only option left for the schools and colleges to conduct the classes. Schools and colleges adopted various apps for teaching purposes. The only necessity for online class is a device and a good internet connection. But the problem is today also there are so many areas where there is not good network connectivity or no internet at all. Moreover, there are so many people who cannot afford a smart device. Seminars, Conferences became online across the globe. Webinars replaced seminars, video conferencing replaced conferences again with the necessity of a smart device and a high-speed network connection. The way of communication was also digitized. Instead of meeting personally, people preferred video calling, chatting. It was a great support for those people who were in isolation or in quarantine. This pandemic gave a rise to e-commerce. People were not going outdoors and were doing online shopping. For example, restaurants that are closed due to lockdowns started online delivery services.

Pandemics are also responsible for a long-term modification in people's choice of food-choices. After the pandemic caused by avian influenza, the intake of non-vegetarian food products drops by almost 80% in the market-place of China [22, 23], which in turn affected the source of income of farm helpers. Pandemics also leave an impact on the mental and physical health of people in various ways and changes the way of living. One more case of such pandemics is Zika-virus which left a generation with neurological disorders; the same disorder was detected in new-born children in Brazil that has imposed critical lifelong boundaries [24]. It has happened many times when pandemics lead to an increase in the death rate. Similarly, the 'Black Death' pandemic in fourteenth-century destroyed almost half of the population of Europe [25]. These types of cases were more in the twentieth century, three main pandemics were Spanish-flu (1919–1920), this pandemic caused around 40-million losses of life [26]; Asian-flu pandemic (1957–1958) killed almost 2-million inhabitants, Hong Kong-flu pandemic (1968–1969) killed approx. 1-million population [27]. And then SARS-pandemic outbreak (2003), 8000 plus infected cases were reported with 700 plus deaths all over the world in just some 6–7 months (approx. 9%) [28].

11.6.2 EFFECTS ON ECONOMIES

During a pandemic, schools, industries, shops, airports, railway stations, and other income sources of people were closed for months which affect the economy of worldwide. Many people lost their jobs or paid half salary. Many workers have lost their jobs as industries are closed for months. Moreover, pandemics affect many businesses, for example, tourism, hotels, etc., severely. According to the news report, 36-million people have already applied for unemployment reimbursements since March which can also be termed as a quarter of the employable age people. The tourism has been badly spoiled, due to lockdown, many flights were canceled even people were restricted to travel from one place to another to stop the virus.

In 2020 whole world, is facing COVID-19. This pandemic first sensed in China and has now infected 188 countries. During this pandemic government have two concerns first, to save people from the pandemic, and second to maintain the economy. The stock market is trading lower for 5–6 months. This pandemic shattered the economy of each country. Almost every country's GDP (gross domestic product) is going down to an extent that it is very difficult to estimate how much time it will take to recover. In many countries, for example, UK reduced interest rates, to encourage people to borrow so that GDP can be boosted. It has been seen a rise in unemployment across the world. According to IMF, 29 June 2020, in Italy unemployment rate has risen from 10 to 12.7% and in the United Kingdom, it has risen from 3.8 to 4.8%. Moreover, there has been noted a fall in marketing as people were less interested in going out from homes due to disease. According to IMF (International Monetary Fund) the expected slowdown in global economy is 3% that is sharpest slow down since 1930s. According to the IMF's estimation, the worldwide economy is at its lowest with a growth rate of approx. 3% this year after coronavirus outbreak which is "far worse" than global fiscal crises in 2009. Even the biggest economies like the USA, UK, Japan, France, Germany, Spain, and Italy are predicted to have the least growth rates this year by 5.9, 6.5, 5.2, 7.2, 7, 8, and 9.1%, respectively. Developed economies are having more difficulties than developing economies, the total growth rate of all developed economies has been predicted only- 6% this year. While developing economies are predicted to have a contract of- 1%. If China alone is not included in this list of countries, the total growth rate of this year would be only about- 2.2%. Chinese economy fell by 36% only in the first few months of coronavirus outbreak, whereas South Korean economy dropped by 5.5%. These are the countries that did not follow the rule of imposing the lockdown but opted out of the approach of forceful testing, tracking the infected cases and isolating.

The GDP of Europe fell by 19.3% approximately just because of decrease in travel and tourism. The fall in oil prices during March month of 2020 because of the decrease in transportation was hit during the lockdown phase in several countries simultaneously, transportation alone demands for 60% of the oil produced worldwide. This is not a case only with oil industry in fact in the first few months of this year in China; the fell in natural gas demand is also high due to coronavirus suppression measures consequently many buyers terminated their imports. When China, USA, and Europe imposed lockdown the sale of industrial metals is also decreased because of the factories closed down. According to the IMF, China is the only country that demands approximately half of the global consumption of industrial metals. IMF has also declared a decrease in the price of food by 2.6% approx. in 2020, due to disturbance in supply chain, delays at borders, export restrictions, food-security issues in coronavirus affected areas.

In the lockdown phase, when the price of many cereals, fruits, Arabica coffee, and seafood has increased, at the same time, the prices of meat, tea, cotton, and wool have declined. Additionally, the drop in the price of oil also put a descending pressure on palm oil prices, sugar prices, soy oil and corn prices. This COVID-19 affected the Indian economy in various ways [29]. It affected chemical industry, auto industry, electronic industry due to the unavailability of raw material due to the lockdown and borders were sealed. Even poultry industry is not untouched with this disease. There were rumors that CoVs transmitted through consumption of chicken; as a result, people stopped buying meat, fish, chicken, eggs. Due to the less demand, the price of chicken is dropped by almost 70%. However, it is important to note here that the consequences of pandemics are not even in every sector. There are some industries which are profited during these pandemics, for example, pharmaceutical companies involved in producing vaccines, medicines, sanitizer, masks, or any other product helpful in breaking the chain of virus.

KEYWORDS

- **AI for pandemic detection**
- **big data for future pandemic prediction**
- **internet of things**
- **pandemic data analysis**
- **pandemic detection**
- **post-pandemic effects**
- **smart technologies**

REFERENCES

1. Rama, K. R. K., (2020). smart technologies for fighting pandemics: The techno and human-driven approaches in controlling the virus transmission. *Government Information Quarterly, 37*(3), 101481. published by Elsevier inc.
2. Jordà, Ò., Sanjay, R. S., & Alan, M. T., (2020). *Longer-Run Economic Consequences of Pandemics*. Federal Reserve Bank of San Francisco, Working Paper 2020-09. [Online] https://doi.org/10.24148/wp2020-09.
3. *Deep Analysis of Global Pandemic Data Reveals Important Insights by Margaretta Colangelo*, (2020). Cofounder and managing partner at Deep Knowledge Group, Forbes.   https://www.forbes.com/sites/cognitiveworld/2020/04/13/covid-19-complexity-demands-sophisticated-analytics-deep-analysis-of-global-pandemic-data-reveals-important-insights/?sh=3d4fd5ba2f6e (accessed on 22 June 2021).
4. Ahmed, B., (2016). *IoT Standardization and Implementation Challenges*. IEEE internet of things, newsletter. https://www.bbvaopenmind.com/en/technology/digital-world/iot-standardization-and-implementation-challenges/ (accessed on 22 June 2021).
5. Matthew, D. B., David, G. G., Osonde, A. O., Andrew, M. P., Ricardo, S., & Claude, M. S., (2020). *Don't Make the Pandemic Worse with Poor Data Analysis*. by RAND blog.   https://www.rand.org/blog/2020/05/dont-make-the-pandemic-worse-with-poor-data-analysis.html (accessed on 22 June 2021).
6. Polonsky, J. A., et al., (2019). Outbreak analytics: A developing data science for informing the response to emerging pathogens. *Phil. Trans. R. Soc. B, 374*, 20180276. http://dx.doi.org/10.1098/rstb.2018.0276.
7. Hadley, W., (2014). Tidy data. *Journal of Statistical Software, 59*(10), 1–23. published by the foundation for open access statistics, ISSN: 1548-7660. (doi: 10.18637/jss.v059.i10).
8. Anne, C., et al., (2017). Key data for outbreak evaluation: Building on the Ebola experience. *Phil. Trans. R. Soc. B, 372*(1721), 20160371. (doi: 10.1098/rstb.2016.0371).
9. Macharia, P., Dunbar, M. D., Sambai, B., Abuna, F., et al., (2015). Enhancing data security in open data kit as an mHealth application. *International Conference on Computing, Communication and Security (ICCCS)*. Pamplemousses, Mauritius. (doi: 10.1109/cccs.2015.7374205).
10. Institute for Health Metrics and Evaluation (IHME), (2018). *Diarrhea*. Seattle, WA: IHME, University of Washington. Available from: https://vizhub.healthdata.org/lbd/diarrhoea (accessed on 14 July 2021).
11. Farrington, C. P., Andrews, N. J., et al., (1996). A statistical algorithm for the early detection of outbreaks of infectious disease. *Journal of the Royal Statistical Society, Series-A (Statistics in Society)* (Vol. 159, No. 3, pp. 547–563). Published by Wiley. (doi: 10.2307/2983331).
12. Jetsada, A., Yunyong, P., & Yodchanan, W., (2011). *IEEE: Wireless Sensor Network-based Smart Room System for Healthcare Monitoring*. 978-1-4577-2138-0/11/$26.00 © 2011 IEEE.

13. Zheng, L., (2006). ZigBee wireless sensor network in industrial applications. *SICE-ICASE International Joint Conference* (pp. 1067–1070). Busan. doi: 10.1109/SICE.2006.315751.

14. Murat, D., & Cevat, B., (2015). Smart technologies with wireless sensor networks. *Social and Behavioral Sciences 195*(2015), 1915–1921. Science direct 1877-0428 © 2015, published by Elsevier Ltd. http://creativecommons.org/licenses/by-nc-nd/4.0/ (accessed on 22 June 2021).

15. Krishna, K., (2020). *The World Wasn't Prepared for COVID-19: In Future, AI Will Curb the Next Pandemic.* "The economic times." https://economictimes.indiatimes.com/magazines/panache/the-world-wasnt-prepared-for-covid-19-in-future-ai-will-curb-the-next-pandemic/articleshow/74776209.cms?from=mdr (accessed on 22 June 2021).

16. Zhu, H., Podesva, P., Liu, X., Zhang, H., Teply, T., Xu, Y., Chang, H., et al., (2020). IoT PCR for pandemic disease detection and its spread monitoring. *Sensors and Actuators B Chemical, 303,* 127098. https://doi.org/10.1016/j.snb.2019.127098.

17. Wilson, K., & Brownstein, J. S., (2009). Early detection of disease outbreaks using the internet. *CMAJ: Canadian Medical Association Journal, 180*(8), 829–831. https://doi.org/10.1503/cmaj.090215.

18. Huang, J., Zhao, H., & Zhang, J., (2013). Detecting flu transmission by social sensor in China. *IEEE International Conference on Green Computing and Communications and IEEE Internet of Things and IEEE Cyber, Physical and Social Computing* (pp. 1242–1247). Beijing. https://doi.org/10.1109/GreenCom-iThings-CPSCom.2013.216.

19. Turcu, C. E., & Turcu, C. O., (2013) Internet of Things as key enabler for sustainable healthcare delivery. *Procedia - Social and Behavioral Sciences, 73*(27), 251–256 (2013).

20. Mathew, A., Amreen, S. A. F., Pooja, H. N., & Verma, A., (2015). Smart disease surveillance based on internet of things (IoT). *International Journal of Advanced Research in Computer and Communication Engineering, 4*(5), 180–183.

21. Sareen, S., Sood, S. K., & Gupta, S. K., (). Secure internet of things-based cloud framework to control zika virus outbreak. *International Journal of Technology Assessment in Health Care, 33*(1), 11–18 (2017).

22. Chen, P., Xie, J. F., Lin, Q., et al., (2019). A study of the relationship between human infection with avian influenza a (H5N6) and environmental avian influenza viruses in Fujian, China. *BMC Infect. Dis., 19,* 762. https://doi.org/10.1186/s12879-019-4145-6.

23. Qiu, W., Rutherford, S., Mao, A., & Chu, C., (2017). *The Pandemic and its Impacts, 9, 10,* 2161–6590. | ISSN [online]. doi: 10.5195/hcs.2017.221.

24. Ribeiro, G. S., & Kitron, U., (2016). Zika virus pandemic: A human and public health crisis. *Revista da Sociedade Brasileira de Medicina Tropical, 49*(1), 1–3.

25. Ross, A. G. P., Ross, A. G. P., Olveda, R. M., & Yuesheng, L., (2014). Are we ready for a global pandemic of Ebola virus? *International Journal of Infectious Diseases, 28,* 217, 218.

26. Taubenberger, J. K., & Morens, D. M., (2009) Pandemic influenza - including a risk assessment of H5N1. *Revue Scientifique Et Technique-Office International Des Epizooties, 28*(1), 187–202.

27. Landis, M., (2007). Pandemic influenza: A review. *Population and Development Review, 33*(3), 429–451.

28. Wong, G. W., & Leung, T. F., (2007). Bird flu: Lessons from SARS. *Pediatric Respiratory Reviews, 8*(2), 171–176. doi: 10.1016/j.prrv.2007.04.003.

29. Sunil, K., Pratibha, B. T., & Pandurang, A. K., (2020). *Impact of Coronavirus (COVID-19) on Indian Economy* (Vol. 2, No. 4). In Agriculture & food: E-newsletter.

30. Ahmed, E., et al., (2017). The role of big data analytics in the internet of things. *Computer Networks.* http://dx.doi.org/10.1016/j.comnet.2017.06.013.

31. Bashir, M. R., & Gill, A. Q., (2016). Towards an IoT big data analytics framework: Smart buildings systems. In: *IEEE 18th International Conference on High Performance Computing and Communications, IEEE 14th International Conference on Smart City, IEEE 2nd International Conference on Data Science and Systems (HPCC/SmartCity/DSS).*

CHAPTER 12

Detection of Emotional Cues of Depression Due to COVID-19 Pandemic

ABHISHEK A. VICHARE[1] and SATISHKUMAR VARMA[2]

[1]*PhD Scholar, Department of Computer Engineering, Pillai College of Engineering (Affiliated to the University of Mumbai), New Panvel, Navi Mumbai, Maharashtra – 410206, India, E-mail: vichare1@gmail.com*

[2]*Professor, HOD of Information Technology Department, Pillai College of Engineering (Affiliated to the University of Mumbai), New Panvel, Navi Mumbai, Maharashtra – 410206, India, E-mail: vsat2k@mes.ac.in*

ABSTRACT

Due to the COVID-19 pandemic, many governments across the globe including India declared restrictions on movement of population by announcement of lockdown as a precaution to prevent spread of virus. COVID-19 pandemic and related restrictions have created many issues such as fear of losing job, closed businesses, not able to meet family members or loved ones trapped in other countries or cities due to travel ban, salary cut/deductions, no social life (closed restaurants, pubs, hotels) as well unavailability of alcoholic beverages. This has impacted on psychological behavior and mental health of many.

A number of symptoms of depression as well cues of ill mental health are visible in the general population. Concerned issues of population in India are fear of job security, fear of COVID-19 (corona) infection, fear of economic instability and recession, anxiety, stress along with other problems caused by COVID-19 pandemic. In this research work, depression cues are analyzed from the data collected by online survey. This survey was conducted in lockdown period (May 2020). Questionnaire of survey was based on instructions in generalized anxiety disorder scale (GAD-7) tool along with questionnaire from patient health questionnaire (PHQ-9). The objective of this research is to identify the effect of COVID-19 pandemic on emotion and behavioral depressing indicators.

12.1   INTRODUCTION

Coronavirus (CoVs) was initially recorded in Wuhan city (capital of Hubei province in China) in the month of December 2019. Wuhan city is considered as a center of an COVID-19 pandemic [1]. The virus is known as SARS-COV-2 and ailments result of this influenza is referred to as "coronavirus disease 2019," which is also known as COVID-19. In February, WHO (world health organization) declared the name to this virus as a COVID-19. Since then, this disease is spreading across the globe in a terrific manner. WHO announced COVID-19 as a health emergency at the global level on 30th January 2020. Common signs of COVID-19 are high fever, cough, and tiredness. Symptoms also include diarrhea, headache, and loss of smell or test [1].

On 10th May 2020, a total number of COVID-19 cases in India was 41472, with 2109 death and 19357 cured/discharged as per government of India report. At global level total cases reported for COVID-19 are 3,862,676 with deaths 265,961 (on 9th May 2020) as per the report by WHO. A total number of CoVs cases today (on 24th September 2020) in India are 5,737,197 and 91,204 deaths. A total number of CoVs cases today (on 24th September 2020) at global level reported for COVID-19 are 32,134,999 and 982,698 deaths. Medicine/vaccine for COVID-19 is not available till date [2]. Therefore, prevention is only cure considered at this stage in India and world widely.

Indian government declared first domestic lockdown from 25th March to 14th April of year 2020. Second lockdown was a continuation to the first lockdown from 15th April to 3rd May of year 2020. Government of India again extended the lockdown by two weeks until 17th May 2020.

The Government of India prohibited the following activities in this lockdown phase:

- Domestic as well international air travel;
- Railway, bus, and metro services;
- All educational campuses and training facilities;
- Industrial and commercial activities except work from home facility;

- Entertainment activities including malls, cinema halls, gyms, swimming pools, parks;
- All public gathering events as well religious places.

After 1st May, the government has declared relaxation in some areas based on red, green, and orange zone. These restrictions as well many fear factors associated with COVID-19 pandemic have temporary and prolonged effects on behavioral health. Because of many threats associated including how normalcy will return, period of lock down is challenging for many people across nation with different age groups and background. Stress during this pandemic outbreak can create impact on sleep patterns, eating patterns, concentration on work or study, behavioral routine. These conditions may worsen the mental health for those who are already patients of depression or mental health. People in a high-risk category such as older people, patients related to diabetics, kidney or high blood pressure may react strongly in such stressful environment.

The mental consequences of the strict lockdown measures implemented by governments worldwide to fight the current COVID-19 pandemic are currently unknown [3]. There is a need to access a state of mind of front-line workers such as medical health practitioners, support medical staff and first respondent in this pandemic. COVID-19 has created a serious impact on mental health across the globe. Many behavioral issues related to anxiety and stress like helplessness, fear; has been noticed among individuals as well community levels [1]. Early detection of behavioral disorders can help for intervention strategies effectively.

This chapter has total five: Section 12.1: "Introduction;" Section 12.2: "Need of Emotion Analysis;" Section 12.3: "Online Survey;" Section 12.4: "Conclusion;" and Section 12.5: "Future Scope." This is possibly the first chapter where emotions related to psychological issues are studied with Indian perspective for COVID-19 epidemic situation by applying the GAD-7 tool and PHQ-9 method.

12.2 NEED OF EMOTION ANALYSIS

12.2.1 CURRENT ISSUES

Growth of global economy is affected badly. Sectors like aviation, tourism, and hospitality expects very low growth in upcoming years. COVID-19 scenario is a humanitarian challenge and economic as well social consequences also need to looked at [4].

Eruption of COVID-19 has created threat of widespread panic and anxiety and health issues in public at global level [2, 5, 6]. Patients with ill mental health are vulnerable to infectious diseases [6]. Mental Health of Chinese citizens is assessed in the COVID-19 pandemic in China through an online survey. The study suggested a high risk of mental illness in China due to COVID-19 [7]. Coronavirus has created fear, stress as well mental disorders. According to the Center for Disease Control (CDC) and Prevention, COVID-19 period will be stressful for humans [8]. There is a high risk of serious impact on mental health on elderly people and people with chronic health issues [9, 29].

Health practices like isolation and quarantine of people are used by the Indian government to prevent the spread of disease. These practices impose change in daily routine and exploit emotional vulnerabilities [10].

COVID-19 has resulted in serious psychological consequences in a huge number. Very low and difficult predictability and uncertainty of COVID-19 are also reasons for ill mental health [11]. COVID-19 is an extremely contagious virus with a lengthy time to symptoms get noticeable as well develop; similar to SARS-CoV-2 (severe acute respiratory syndrome coronavirus 2) [12].

WHO and other agencies world widely has focused on COVID-19 and precautions such as social distancing and isolation. But these measures do not consider the impact on mental health of public associated [13].

The requirement of a robust mental well-being system in cases of health emergencies was mentioned by author Duan and Zhu in their research paper [14]. Anxious depression is major distressing disorder (MDD) accompanying another issue of panic as well restlessness [15]. A person with depression needs extra care; otherwise, it can create life threat as well serious damage to mental well-being. Considering these facts and literature, it is very clear that there is a strong need to analyze human emotions on the basis of COVID-19 pandemic.

For some parts of China, Wang et al. conducted research on emotional status and its associated aspects. Study population (participants) was 600. The self-rating anxiety scale (SAS) as well the self-rating depression scale (SDS) method were applied in online survey questionnaire. Percentage of anxiety and depression reported was 6.33 and 17.17 respectively. It was concluded that there is a need to pay attention towards public psychological state [16].

Wang et al. conducted an online survey to determine psychological disorders among Chinese adolescents in COVID-19 pandemic. PHQ-9 and GAD-7 method is used to prepare questionnaire. Study population (participants) was 8079. Percentage of depression indicators were 43.7% and anxiety cues found in 37.4% of participants [17, 31].

Sigdel et al. conducted a study in Nepal with the general population. Survey was conducted on social media through questionnaire using PHQ-9 along with GAD-7 method. Target population (participants) was 349. Percentage of depression and anxiety was 34.0% and 31.0% [18].

Agberotimi et al. conducted a survey to understand psychological issues in COVID-19 situation of medical personnel along with the general population of Nigeria [19]. Effect of event scale-revised (IES-R), GAD-7, PHQ-9 along with Insomnia Severity Index were used in questionnaire. Study population (participants) was 884. Percentage of distress as well as anxiety was 23.5% and 49.6% sequentially.

12.2.2   IMPORTANCE OF PSYCHOLOGICAL BEHAVIOR STUDY

Stress can be conceptualized as being confronted with a challenging situation that requires some kind of adaptation [20]. Human being can handle the stress as well tension up to certain levels and phases. But sometimes this can create serious behavioral, suicidal, depressive, and mental issues in some humans [21]. E. A. Holmes et al. expressed concerned about psychological consequences due to COVID-19 to plan way for public health. Survey was done in U.K. Need of collecting data for mental health was highlighted due to COVID-19 scenario in UK [22].

Human body consist behavioral immune system (BIS) which is bunch of emotional processes which encourages human body to avoid diseases [23]. Due to BIS humans develop negative emotions like anxiety, aversion [11]. In the long term, negative emotion can create serious impact on immune system and physiological Mechanisms [24]. Hence, it is necessary to take precautions to avoid psychological emotions created by COVID-19 pandemic. Mental health is and continues to be a prominent plague for the civilized world [7].

Also, the wrong information and rumors about lockdown, government orders have affected on public psychological health [25]. The situation raises concerns of widespread panic in individuals [26]. This situation is created due to a complete lock down in India due to COVID-19 (corona) pandemic.

This situation is challenging for many because of the imminent threat of getting infected, isolation, quarantine, the problems arising from the governmental restrictions and the uncertainty about how and when people will be able to return to their normal daily routine.

12.3   ONLINE SURVEY

12.3.1   RESEARCH METHODOLOGY USED

The questionnaire used in the survey is based on the instruction manual of the patient health questionnaire (PHQ) along with generalized anxiety disorder (GAD-7). At first PHQ-9 was designed to determine GAD-7. Depression is assessed using the PHQ-9 [5]. GAD-7 is widely used to conduct research in anxiety disorders as well in medical practice [27].

PHQ was designed on the basis of the PRIME-MD symptomatic tool for general psychological problems. PHQ also provide facility to respondents to fill the questionnaire by themselves unlike PRIME-MD. PHQ-9 version is considered as reliable and valid method for measurement of identification of depression and depression severity. Because of these quality PHQ-9 is considered a useful clinical and research tool [28, 30].

Creating database about behavioral or emotional state of public with different facet is required on priority for present state of COVID-19 as well any untoward medical emergencies in future. Therefore, we conducted an online survey for more than 200 people living in India to explore emotional characteristics to find out symptomatic cues of depression in coronavirus scenario (COVID-19 pandemic). The objective of this survey was to understand stress and other behavioral patterns of participants in lockdown. We collected data for eight days in lockdown period (May 2020) in India.

12.3.2   DATA COLLECTION PROCEDURE

Data analysis is done on the data collected from an online questionnaire answered by participant facing lockdown in India. This questionnaire was answered at a single time by participant. In the situation of complete lockdown and restrictions on the movement; it is very difficult to conduct a traditional type survey in India to assess the behavior and emotion of people. This is self-reported data by participant. Survey went online on 5th May 2020, which was the third lockdown phase in India. Objective of study, to identify effect of COVID-19 epidemic on emotion as well behavioral depressive symptoms will also help NGO's, government, practitioners (Figures 12.1 and 12.2).

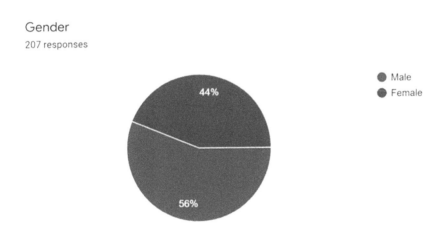

FIGURE 12.1   Participants gender factor.

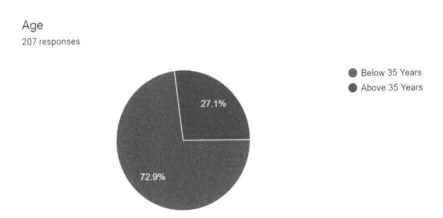

FIGURE 12.2   Participants age factor.

Most of the participants in the survey were youth with age below than 35 years. 56% participants were male. 72.9% participants were in the category of less than 35 years old, and 27.1% of participants were in more than 35 years old category. 39% of participants were students and 41% belongs to private-sector employees (Figure 12.3).

Occupation

207 responses

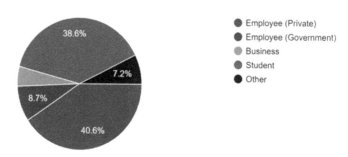

FIGURE 12.3   Categories of participant on the basis of occupation.

Many participants have reported multiple issues due to COVID-19 pandemic, as shown in Figure 12.4.

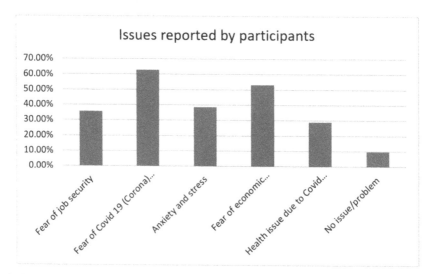

FIGURE 12.4   Issues reported by participant due to COVID-19 pandemic.

12.3.3   IMPACT ON GENERAL POPULATION

After all, emotions are difficult to communicate, but digital self-administered platforms facilitate mechanism to express psychological states into physical form [15, 33].

Around 10% of people reported no issue or problem because of current situation, 36% people are worried about their job security, 63% people are worried about getting infected by COVID-19, 39% people have reported issues of anxiety and stress due to COVID-19 situation. 53% people have fear of economic instability or recession and 29% participant of survey have noted health issues like less sleep, loneliness, restlessness, too worried.

One quarter of participants reported health related problems due to COVID-19. More than half of the participants noted fear of economic instability in lock down report. There are other issues also for which people are concerned like closed school, negligence towards child care and elderly people, social distancing, lack of physical exercise, unavailability of medicals, rescheduling of family functions, no social life (closed movie theater, restaurants).

Guidelines and information through helplines will help to people those who are having emotional suffering. Limited use of social media and news may also help as a factor to reduce the panic and stress. A healthy diet, healthy routine like good sleep [32], spending quality time with loved one's/family members, taking advice from medical practitioners/counselors will help in emotional distress (Table 12.1).

TABLE 12.1   COVID-19 Issues Reported by Survey Participant

Impact of COVID-19	Number of Participant	Percentage (%)
Fear of job security	74	35.70%
Fear of COVID-19 (Corona) infection	130	62.80%
Anxiety and stress	80	38.60%
Fear of economic instability/recession	110	53.10%
Health issue due to COVID-19 (less sleep, loneliness, restlessness, too worried)	60	29%
No issue/problem	20	9.70%

12.3.4 ETHICAL CONSIDERATIONS

Anonymity of all the participants of the survey is maintained. Digital surveys, studies bank on the collaboration with participants and information fetched in this survey is voluntarily self-reported. Timestamps of participant opinion was recorded.

12.4 CONCLUSION

This study was conducted to determine facets associated with the psychological issues of the Indian population at the start of the CoVs epidemic lockdown period. Finally, we explored few emotional behavior issues that have increased a number of cues which leads to depression. This chapter also emphasizes increase in depressive symptoms due to COVID-19 pandemic. This study can help NGO's, government, and practitioners in the medical field.

It is highlighted that psychological support should be provided to enhance psychoneurotic immunity against COVID-19. Until solid data are available, we can only speculate about the actual consequence of COVID-19 on behavioral health. Findings in this study may become reference which can benefit agencies to battle versus COVID-19 pandemic by also focusing on public emotion and to help in the development of policies to policymakers. This chapter has limitations because of evolving nature of COVID-19 and uncertainty over the symptoms.

Government agencies should provide update on COVID-19 information on regular basis to the masses. Information on rumors will also impact positively on people's emotional state. Optimistic use of electronic media and positive attitude towards pandemic can reduce psychological stress. Exhaustive research and analysis of risk factors for individuals and communities need to be done for preventive and cure measurement of the general population due to this pandemic situation.

12.5 FUTURE SCOPE

With the help of government officials or government departments, survey can be conducted for health workers; those who are constantly providing health service to COVID-19 patients. It seems, these workers are having a lot of psychological pressure due to long working hours and huge work-load. Official survey can provide more insights into behavioral issues of healthcare professionals like doctors, nurses, ward boys, and other medical staff. Public emotional state may get changed over period of time; therefore, follow-up study in a similar way will be helpful to understand long-term implications [34].

ACKNOWLEDGMENT

We would like to show gratitude to the management of Mahatma Education Society, Panvel, and Narsee Monjee Institute of Management Studies University for providing guidance and support in this study.

KEYWORDS

- **behavioral immune system**
- **coronavirus**
- **COVID-19**
- **depression**
- **emotion**
- **generalized anxiety disorder**
- **psychological behavior**
- **self-rating anxiety scale**

REFERENCES

1. Salari, N., Hosseinian-Far, A., Jalali, R., Vaisi-Raygani, A., Rasoulpoor, S., Mohammadi, M., Rasoulpoor, S., & Khaledi, B., (2020). Prevalence of stress, anxiety, depression among the general population during the COVID-19 pandemic: A systematic review and meta-analysis." *Globalization and Health, 16*, 1–11. doi: 10.1186/s12992–020–00589-w.
2. Zhou, T., Liu, Q., Yang, Z., Liao, J., Yang, K., Bai, W., Lu, X., & Zhang, W., (2020). Preliminary prediction of the basic reproduction number of the Wuhan novel coronavirus 2019-nCoV," *J. Evid. Based Med., 13*(1), 3–7. PMID: 32048815, PMCID: PMC7167008. doi: 10.1111/jebm.12376.

3. De Quervain, D., Aerni, A., Amini, E., Bentz, D., Coynel, D., Gerhards, C., Fehlmann, B., et al., (2020). *The Swiss Corona Stress Study*. doi: 10.31219/osf.io/jqw6a.

4. Hua, J., & Shaw, R., (2020). Coronavirus (COVID-19) "infodemic" and emerging issues through a data lens: The case of China. *Int. J. Environ. Res. Public Health, 17*(7), 2309. doi: 10.3390/ijerph17072309.

5. Nguyen, H. C., Nguyen, M. H., Do, B. N., Tran, C. Q., Nguyen, T. T. P., Pham, K. M., Pham, L. V., et al., (2020). People with suspected COVID-19 symptoms were more likely depressed and had lower health-related quality of life: The potential benefit of health literacy. *J. CCli.nMed., 9*(4), 965. PMID: 32244415, PMCID: PMC7231234. doi: 10.3390/jcm9040965.

6. Kim, S., & Kuan-Pin, S., (2020). Using psychoneuroimmunity against COVID-19. *Brain, Behavior, and Immunity*. doi: 10.1016/j.bbi.2020.03.025.

7. Huang, Y., & Zhao, N., (2020). Chinese mental health burden during the COVID-19 pandemic. *Asian J. Psychiatr., 51*, 102052. PMID: 32361387; PMCID: PMC7195325. doi: 10.1016/j.ajp.2020.102052.

8. Singh, J., & Singh, J., (2020). COVID-19 and its impact on society. *Electronic Research Journal of Social Sciences and Humanities, 2*(I).

9. Amerio, A., Aguglia, A., Odone, A., Gianfredi, V., Serafini, G., Signorelli, C., & Amore, M., (2020). COVID-19 pandemic impact on mental health of vulnerable populations. *Acta Bio-Medica: Atenei Parmensis, 91*(9-S), 95–96.

10. Campos, Juliana, A. D. B., Martins, B. G., Campos, L. A., Marôco, J., Saadiq, R. A., & Ruano, R., (2020). Early psychological impact of the COVID-19 pandemic in Brazil: A national survey. *J. Clin. Med., 9*(9), 2976.

11. Brooks, S., Webster, R., Smith, L., Woodland, L., Wessely, S., Greenberg, N., & Rubin, G., (2020). The psychological impact of quarantine and how to reduce it: Rapid review of the evidence. *The Lancet, 395*. doi: 10.1016/S0140-6736(20)30460-8.

12. Li, S., Wang, Y., Xue, J., Zhao, N., & Zhu, T., (2020). The impact of COVID-19 epidemic declaration on psychological consequences: A study on active Weibo users. *Int. J. Environ. Res. Public Health, 17*(6), 2032. PMID: 32204411; PMCID: PMC7143846. doi: 10.3390/ijerph17062032.

13. Shah, K., Kamrai, D., Mekala, H., Mann, B., Desai, K., & Patel, R. S., (2020). Focus on mental health during the coronavirus (COVID-19) Pandemic: Applying learnings from the past outbreaks. *Cureus, 12*(3), e7405. PMID: 32337131, PMCID: PMC7182052. doi: 10.7759/cureus.7405.

14. Duan, L., & Zhu, G., (2020). Psychological interventions for people affected by the COVID-19 epidemic. *Lancet Psychiatry, 7*(4), 300–302. doi: 10.1016/S2215-0366(20)30073-0.

15. Nadeem, M., (2016). *Identifying Depression on Twitter*. arXiv:1607.07384.

16. Wang, Y., Di, Y., Ye, J., & Wei, W., (2020). Study on the public psychological states and its related factors during the outbreak of coronavirus disease 2019 (COVID-19) in some regions of China. *Psychol Health Med., 30*, 1–10. PMID: 32223317. doi: 10.1080/13548506.2020.1746817, Epub ahead of print.

17. Zhou, S. J., Zhang, L. G., Wang, L. L., Guo, Z. C., Wang, J. Q., Chen, J. C., Liu, M., et al., (2020). Prevalence and socio-demographic correlates of psychological health problems in Chinese adolescents during the outbreak of COVID-19. *Eur. Child*

Adolesc. Psychiatry, 29(6), 749–758. PMID: 32363492; PMCID: PMC7196181. doi: 10.1007/s00787-020-01541-4.

18. Sigdel, A., Bista, A., Bhattarai, N., Poon, B. C., Giri, G., & Marqusee, H., (2020). *Depression, Anxiety and Depression Anxiety Comorbidity Amid COVID-19 Pandemic: An Online Survey Conducted During Lockdown in Nepal.* medRxiv. doi: 10.1101/2020.04.30.20086926.

19. Agberotimi, S. F., Akinsola, O., Oguntayo, R., & Olaseni, A., (2020). Interactions between socioeconomic status and mental health outcomes in the Nigerian context amid COVID-19 pandemic: A comparative study. *Frontiers in Psychology.*

20. McEwen, B. S., (1998). Stress, adaptation, and disease. Allostasis and allostatic load. *Ann. N. Y. Acad. Sci., 840*, 33–44. PMID: 9629234. doi: 10.1111/j.1749-6632.1998. tb09546.x.

21. De Kloet, E. R., Joels, M., & Holsboer, F., (2005). Stress and the brain: From adaptation to disease. *Nat. Rev. Neurosci., 6*, 463–475. doi: 10.1038/nrn1683.

22. Holmes, E. A., O'Connor, R. C., Perry, V. H., Tracey, I., Wessely, S., Arseneault, L., Ballard, C., et al., (2020). Multidisciplinary research priorities for the COVID-19 pandemic: A call for action for mental health science. *Lancet Psychiatry, 7*(6), 547. doi: 10.1016/S2215-0366(20)30168–1.

23. Huang, C., Wang, Y., Li, X., Ren, L., Zhao, J., Hu, Y., Zhang, L., et al., (2020). Clinical features of patients infected with 2019 novel coronavirus in Wuhan, China. *Lancet, 395*(10223), 497–506. PMID: 31986264; PMCID: PMC7159299. doi: 10.1016/ S0140-6736(20)30183-5.

24. Terrizzi, Jr. J., Shook, N., & Mcdaniel, M., (2012). The behavioral immune system and social conservatism: A meta-analysis. *Evolution and Human Behavior, 34*. doi: 10.1016/ j.evolhumbehav.2012.10.003.

25. Bao, Y., Sun, Y., Meng, S., Shi, J., & Lu, L., (2020). 2019-nCoV epidemic: Address mental health care to empower society. *Lancet (London, England), 395*, 10224. doi: 10.1016/s0140-6736(20)30309-3.

26. Kiecolt-Glaser, J. K., McGuire, L., Robles, T. F., & Glaser, R., (2002). Emotions, morbidity, and mortality: New perspectives from psychoneuroimmunology. *Annu Rev Psychol., 53*, 83–107, PMID: 11752480. doi: 10.1146/annurev. psych.53.100901.135217.

27. Rutter, L. A., & Brown, T. A., (2017). Psychometric properties of the generalized anxiety disorder scale-7 (GAD-7) in outpatients with anxiety and mood disorders. *J. Psychopathol. Behav. Assess, 39*, 140–146. doi: 10.1007/s10862-016-9571-9.

28. Kroenke, K., Spitzer, R. L., & Williams, J. B., (2001). The PHQ-9: Validity of a brief depression severity measure. *J. Gen. Intern. Med., 16*(9), 606–613. doi: 10.1046/j.1525-1497.2001.016009606.x. PMID: 11556941; PMCID: PMC1495268.

29. Serafini, G., Parmigiani, B., Amerio, A., Aguglia, A., Sher, L., & Amore, M., (2020). The psychological impact of COVID-19 on the mental health in the general population. *QJM: An International Journal of Medicine, 113*(8), 531–537.

30. Gwatkin, R. B. L., (1989/1990). The technical writer's handbook by Matt Young University Science Books Mill Valley California. *Mol. Reprod. Dev., 25, 232*(1), 22–50, 97.

31. Zhou, F., Yu, T., Du, R., Fan, G., Liu, Y., Liu, Z., Xiang, J., Wang, Y., et al., (2020). Clinical course and risk factors for mortality of adult inpatients with COVID-19

in Wuhan, China: A retrospective cohort study. *The Lancet, 395*. 10.1016/S0140-6736(20)30566-3.

32. Ernstsen, L., & Havnen, A., (2020). Mental health and sleep disturbances in physically active adults during the COVID-19 lockdown in Norway: Does change in physical activity level matter? *Sleep Medicine*. doi: 10.1016/j.sleep.2020.10.004 [Epub ahead of print].

33. Akshi, K., Aditi, S., & Anshika, A., (2019). Anxious depression prediction in real-time social data. *Accepted for Publication in the Proceeding of International Conference on Advanced Engineering, Science, Management and Technology (ICAESMT19).*

34. Young, M., (1989). *The Technical Writer's Handbook*. Mill Valley, CA: University Science.

Index